LEARNING-BASED SOFT SENSING AND PREDICTIONS FOR PROCESS INDUSTRIES

LEARNING-BASED SOFT SENSING AND PREDICTIONS FOR PROCESS INDUSTRIES

Theory, Methodology and Applications

YONGXIANG LEI
School of Automation
Central South University
Changsha, China

HAMID REZA KARIMI
Department of Mechanical Engineering
Politecnico di Milano
Milan, Italy

ELSEVIER

ACADEMIC PRESS
An imprint of Elsevier

Academic Press is an imprint of Elsevier
125 London Wall, London EC2Y 5AS, United Kingdom
50 Hampshire Street, 5th Floor, Cambridge, MA 02139, United States

ISBN: 978-0-443-36759-5

For information on all Academic Press publications
visit our website at https://www.elsevier.com/books-and-journals

Publisher: Matthew Deans
Acquisitions Editor: Glyn Jones
Editorial Project Manager: Tripti Rawat
Production Project Manager: Fizza Fathima
Cover Designer: Bakyalakshmi S

Typeset by VTeX

Contents

PART 3 Applications

Preface

With the rapid development of Industry 4.0, artificial intelligence, and big data technologies, modern process industries are undergoing a profound transformation. In complex industrial environments, such as aluminum electrolysis, paste-filling systems, and metallurgical processes, many key quality variables are difficult, expensive, or even impossible to measure directly due to harsh operating conditions, strong coupling effects, large delays, and nonlinear dynamics. These challenges have driven the emergence and rapid advancement of learning-based prediction and soft sensing techniques.

Soft sensing and data-driven prediction provide a powerful alternative to traditional hardware sensors by developing computational models that infer critical process variables from easily measurable signals. Over the past decade, machine learning—especially deep learning—has significantly enhanced the modeling of nonlinear, high-dimensional, and time-dependent industrial systems. However, despite numerous research advances, there remains a clear need for a systematic and structured presentation that bridges theory, methodology, and real industrial applications. This book provides such a comprehensive framework.

The core objective of this book is to present the theory, methodologies, and practical implementations of learning-based prediction and soft sensing in process industries. It integrates classical statistical modeling, modern machine learning techniques, probabilistic forecasting, interpretable modeling, and hybrid frameworks into a unified perspective. Special attention is given to real industrial case studies, including aluminum electrolysis and paste-filling systems, ensuring that theoretical developments are closely connected to practical engineering challenges.

The book is organized into three main parts.

Part 1 – Theory

Part 1 introduces the fundamental concepts of soft sensing and data-driven prediction. It discusses single-variable and multi-variable soft sensing, time series forecasting strategies, recursive and direct prediction methods, and multiple-horizon forecasting. Two representative industrial processes—aluminum electrolysis and paste-filling systems—are presented to illustrate the practical prediction and soft sensing challenges in real-world scenarios.

Part 2 – Methodology

Part 2 focuses on machine learning methodologies for prediction and soft sensing. It systematically introduces deterministic models (e.g., ARIMA, ANN, LSTM, CNN, Transformer, GNN, PINN), probabilistic and uncertainty-aware models (e.g., Bayesian neural networks, quantile regression), interpretable models (e.g., KAN), and hybrid approaches. Both classical techniques and recent advances in deep learning-based time series forecasting are discussed to provide readers with a broad and forward-looking methodological foundation.

Part 3 – Applications

Part 3 presents a series of application-oriented frameworks derived from real industrial problems. These include semi-supervised soft sensors, self-supervised prediction frameworks, probabilistic forecasting models, and hybrid deep learning architectures. Each framework is accompanied by problem formulation, methodology, and performance evaluation, demonstrating how theoretical models can be translated into deployable industrial solutions.

This book is intended for graduate students, researchers, and industrial practitioners in mechanical engineering, control engineering, process engineering, and applied artificial intelligence. It may serve both as a textbook for advanced courses on industrial AI and as a reference for engineers seeking to implement data-driven monitoring and prediction systems in practice.

We hope this book will advance intelligent monitoring, predictive analytics, and data-centric decision-making in process industries and inspire further research at the intersection of machine learning and industrial systems.

Yongxiang Lei and Hamid Reza Karimi
Milan, Italy
February 2026

Acknowledgments

The completion of this book would not have been possible without the support and contributions of many individuals and institutions.

First and foremost, we would like to express our sincere gratitude to our colleagues, collaborators, and students who have contributed to the research works forming the foundation of this book. Many of the methodologies and frameworks presented here are the result of years of collaborative research, discussions, and experimental validation.

We gratefully acknowledge the support from our affiliated institutions, Politecnico di Milano and Central South University, which have provided an inspiring academic environment and strong research infrastructure. The interdisciplinary collaboration between control engineering, machine learning, and industrial process engineering has been essential in shaping the ideas presented in this book.

We also thank the industrial partners who provided valuable datasets, engineering insights, and practical feedback. Their real-world challenges and requirements have guided us in developing solutions that are not only theoretically sound but also practically meaningful.

Financial support from national and international research funding agencies (CSC under Grant No. 202006370101) is sincerely appreciated. Their support has enabled the exploration of advanced learning-based methodologies and their applications in complex industrial systems.

Finally, we extend our heartfelt thanks to our families for their continuous encouragement, patience, and understanding throughout the writing process.

PART 1

Theory

CHAPTER 1

Introduction to soft sensing

1.1 Overview of single-variable soft sensing

Single-variable soft sensing focuses on estimating a specific critical parameter, such as the viscosity of the polymer in chemical plants or the alkalinity in anaerobic digesters. In modern industrial processes [1], the pursuit of greater efficiency [2], better quality control, and a minimized environmental impact have driven the integration of advanced technologies into traditional manufacturing practices. Central to this evolution is the concept of soft sensing, a dynamic approach that amalgamates traditional process measurements with computational modeling and data-driven techniques. This novel paradigm, often referred to as "soft sensing", has quickly emerged as an indispensable tool in the realm of process monitoring [3], offering a plethora of benefits that extend beyond the capabilities of traditional sensors.

1.2 Overview of multiple-variable soft sensing

Multiple-variable soft sensing is an advanced approach that leverages computational models and algorithms to estimate and predict critical process variables in industrial systems, especially those that are difficult or expensive to measure directly. By integrating multiple variables and using sophisticated techniques such as machine learning and statistical modeling, soft sensing enhances process monitoring, control, and optimization.

Soft sensing techniques involve using indirect measurements and data-driven models to infer the state of industrial processes. These methods rely on historical data, real-time measurements, and mathematical models to estimate variables that are otherwise challenging to obtain directly. In the context of multiple-variable soft sensing, several input variables are combined to improve the accuracy and reliability of predictions.

1.2.1 Applications

Industrial process monitoring: Multiple-variable soft sensing is widely used in industries such as metallurgy, chemical processing, and manufacturing.

Learning-Based Soft Sensing and Predictions for Process Industries
https://doi.org/10.1016/B978-0-44-336759-5.00008-1

For example, in metallurgy, soft sensing techniques predict the silicon content in blast furnaces, a variable that is difficult to measure directly. Similarly, in dynamic batch processes like penicillin fermentation and injection molding, neural network-based approaches optimize the prediction of key quality parameters.

Quality prediction: Soft sensing models can predict product quality in real-time, enabling proactive adjustments in the production process. For instance, the principal component-based semi-supervised extreme learning machine (PCSELM) model combines labeled and unlabeled data to enhance predictive accuracy [4]. This approach is particularly useful in industries where direct measurement of quality parameters is impractical or costly.

Fault detection and diagnostics: By monitoring multiple variables simultaneously soft sensing can detect anomalies and potential faults in industrial systems [5]. This early detection allows for timely maintenance and reduces downtime [6].

1.2.2 Techniques and methodologies

Statistical methods: Traditional statistical techniques such as principal component analysis (PCA) and partial least squares (PLS) are commonly used in soft sensing [1,7,2]. These methods help in reducing data dimensionality and extracting meaningful features from large datasets.

Machine learning and neural networks: Machine learning algorithms, including neural networks, play a crucial role in multiple-variable soft sensing. For example, echo state networks (ESNs) integrated with autoencoders handle high-dimensional, nonlinear data and improve prediction accuracy [8]. Additionally, just-in-time learning (JITL) strategies enhance the adaptability and accuracy of real-time process control models [9].

Hybrid approaches: Combining different techniques often yields better results [10]. For instance, integrating domain-specific knowledge with data-driven models improves the robustness and reliability of soft sensing [11]. Hybrid models that leverage both statistical and machine learning methods are particularly effective in handling complex industrial processes [12].

1.2.3 Current challenges

Model interpretability: Complex models, especially those based on deep learning, are often difficult to interpret. This opacity can hinder their acceptance and trust in industrial settings [13].

Computational complexity: Real-time data integration and processing demand efficient computational resources. Techniques like JITL, which require real-time updates, can strain computational capabilities.

Data quality and quantity: Soft sensing relies on high-quality data. Limited or noisy data can affect the accuracy of predictions. Data augmentation techniques, such as virtual sample generation, are being explored to address this challenge.

1.2.4 Future trends

Integration of advanced technologies future trends in multiple-variable soft sensing point toward greater integration of advanced technologies such as machine learning, real-time predictive modeling, and Bayesian approaches. These technologies enable more precise and efficient real-time decision-making in industrial processes.

Enhanced model interpretability research focuses on developing methods to improve the interpretability of complex models. This includes techniques such as feature importance analysis and causal inference algorithms.

Real-time optimization with the rise of Industry 4.0: There is an increasing need for real-time optimization of industrial processes. Soft sensing techniques will play a crucial role in providing the necessary data and insights to enable dynamic process adjustments.

Multiple-variable soft sensing offers a powerful solution for enhancing process monitoring, control, and optimization in industrial systems. By leveraging multiple variables and advanced computational techniques, soft sensing improves the accuracy and reliability of predictions. While challenges remain, ongoing research and technological advancements continue to drive improvements in this field.

1.3 Preview of the book

This book is structured as follows. It inherits from the theory, methodology, and application aspects. It helps both the researchers and practitioners to study soft sensing and prediction and supports practitioners in applying these techniques to solve practical problems in industrial settings. **Part 1, Theory**, covers the fundamental concepts of soft sensing and prediction, along with two representative industrial case studies.

Part 1: Theory (Concept) – Chapters 1–4

Chapter 1 introduces the concept of soft sensing, which covers an overview of single-variable soft sensing and multiple-variable soft sensing.

Chapter 2 introduces the concept of prediction with a focus on both single- and multiple-variable prediction techniques. It also discusses the difference of time series forecasting (TSF) and recursive strategy. In addition, the chapter introduces the concept of multiple-horizon prediction, highlighting its relevance in practical applications.

Chapter 3 introduces the aluminum electrolysis process industry and provides an overview of aluminum production, offering insight into the key concepts behind the process. It also discusses the challenges and prediction problems commonly encountered in aluminum electrolysis.

Chapter 4 presents another industrial process, focusing on underground mining. It covers in detail the paste-filling process, providing background information on this specific techniques. The chapter also discusses prediction and soft sensing challenges specific to this industry. Together with the aluminum electrolysis industry discussed previously, these two case studies lay the foundation for the methodologies and applications explored in later chapters.

Part 2: Methodology – Chapter 5

Chapter 5 introduces two categories of ML-based prediction and soft sensing. It covers the deterministic models, probabilistic models, interpretable and explainable models, and hybrid models.

It covers the autoregressive models (ARIMA), artificial neural network (ANN), long short-term memory (LSTM), convolutional neural network (CNN), autoencoder (AE), transformer, temporal convolutional network (TCN), extreme learning machine (ELM), graph neural network (GNN), and Mamba, physics-informed neural network (PINN).

Regarding the probabilistic neural network, Bayesian neural network (BNN) and Kolmogorov–Arnold networks (KANs) are set as examples for interpretable network.

We leave for the reader's own exploration some advanced variants of TSF models, such as RLinear [14], DLinear [15], DeepAR [16], iTransformer [17], Autoformer [18], Crossformer [19], PatchTST [20], TiDE [21], TimesNet [22], FEDformer [23], and S-Mamba [24].

Part 3: Applications – Chapters 6–14

Chapters 6–14 implement various frameworks for industrial applications, based on different publications.

Chapter 6 presents the CNN-LapsELM framework. For more detail, the reader can refer to "A semi-supervised Laplacian extreme learning machine and feature fusion with CNN for industrial superheat identification."

Chapter 7 states the SAE-WELM framework. A reference is "Processes soft modeling based on stacked autoencoders and wavelet extreme learning machine for aluminum plant-wide application."

Chapter 8 gives LHSS-HELM. We refer to "Manifold semi-supervised learning for aluminum electrolysis temperature identification based on regularized hierarchical extreme learning machine."

Chapter 9 presents the SSDLSTM framework. We refer to "A novel self-supervised deep LSTM network for industrial temperature prediction in aluminum processes application."

Chapter 10 implements CNN-SSDLSTM. It is an improved framework of SSDLSTM. We refer the readers to "A self-supervised temporal temperature prediction method based on dilated contrastive learning."

Chapter 11 proposes BLSTM. We refer the reader to "Data-centric probabilistic temperature prediction with quantified uncertainty using Bayesian machine learning for the aluminum electrolysis process."

Chapter 12 explains DDQRF. The reader can refer to "Underflow concentration soft sensing for cone thickener system based on a direct data-driven quantile regression forecasting."

Chapter 13 shows DualLSTM, which is described in "A novel key quality prediction for a hierarchical cone thickener."

Finally, **Chapter 14** describes DE-LSTM. The reader should refer to "Underflow concentration prediction based on improved dual bidirectional LSTM for hierarchical cone thickener system."

1.4 Abbreviations and notation

AE	Auto-encoder
ANN	Artificial neural network
ARMA	Auto-regressive moving average
BLSTM	Bayesian long short-term memory
BiLSTM	Bidirectional long short-term memory
CI	Confidence interval
CNN	Convolutional neural network
CNN-LapsELM	Convolutional neural network Laplacian extreme learning machine

continued on next page

CNN-SSDLSTM	Convolution self-supervised deep long short-term memory
CT	Cell temperature
CTS	Cone thickener system
DDQRF	Direct data-driven quantile regression forecaster
DBN	Deep belief network
DE-LSTM	Deep efficient long short-term memory
DL	Deep learning
DualLSTM	Dual long short-term memory
ELM	Extreme learning machine
ELU	Exponential linear unit
GLU	Gated linear unit
GRN	Gated recurrent network
GRU	Gated recurrent unit
GNN	Graph neural network
HELM	Hierarchical extreme learning machine
KAN	Kolmogorov–Arnold network
KL	Kullback–Leibler
LSTM	Long short-term memory
LHSS-HELM	Laplacian–Hessian semi-supervised hierarchical extreme learning machine
MAE	Mean absolute error
ML	Machine learning
MSE	Mean squared error
NN	Neural network
PCA	Principal component analysis
PINN	Physics-informed neural network
PPCR	Probabilistic principal component regression
PT	Pot temperature
RBF	Radial basis function
RMSE	Root mean squared error
RNN	Recurrent neural network
SAE	Stacked auto-encoder
SAE-WELM	Stacked auto-encoder wavelet extreme learning machine
SD	Superheat degree
SLFN	Single-layer feed-forward network
SSL	Self-supervised learning
SS-ELM	Semi-supervised ELM
SSDLSTM	Self-supervised deep long short-term memory
SVM	Support vector machine
TSF	Time series forecasting [1]

References

[1] Xiaofeng Yuan, Biao Huang, Yalin Wang, Chunhua Yang, Weihua Gui, Deep learning-based feature representation and its application for soft sensor modeling with variable-wise weighted sae, IEEE Transactions on Industrial Informatics 14 (7) (2018) 3235–3243.

[2] Xiaofeng Yuan, Lin Li, Yalin Wang, Nonlinear dynamic soft sensor modeling with supervised long short-term memory network, IEEE Transactions on Industrial Informatics 16 (5) (2019) 3168–3176.

[3] Xiaofeng Yuan, Lin Li, Yuri A.W. Shardt, Yalin Wang, Chunhua Yang, Deep learning with spatiotemporal attention-based LSTM for industrial soft sensor model development, IEEE Transactions on Industrial Electronics 68 (5) (2020) 4404–4414.

[4] Xudong Shi, Qi Kang, Hanqiu Bao, Wangya Huang, Jing An, Principal component-based semi-supervised extreme learning machine for soft sensing, IEEE Transactions on Automation Science and Engineering (2023).

[5] Mitch Serpas, Yunfei Chu, Juergen Hahn, Fault detection approach for systems involving soft sensors, Journal of Loss Prevention in the Process Industries 26 (3) (2013) 443–452.

[6] Rajamani Doraiswami, Maryhelen Stevenson, Chris Diduch, Identification of Physical Systems: Applications to Condition Monitoring, Fault Diagnosis, Soft Sensor and Controller Design, John Wiley & Sons, 2014.

[7] Xiaofeng Yuan, Jiao Zhou, Biao Huang, Yalin Wang, Chunhua Yang, Weihua Gui, Hierarchical quality-relevant feature representation for soft sensor modeling: a novel deep learning strategy, IEEE Transactions on Industrial Informatics 16 (6) (2019) 3721–3730.

[8] Naima Chouikhi, Boudour Ammar, Amir Hussain, Adel M. Alimi, Novel single and multi-layer echo-state recurrent autoencoders for representation learning, Engineering Applications of Artificial Intelligence 114 (2022) 105051.

[9] Zeyu Yang, Zhiqiang Ge, Rethinking the value of just-in-time learning in the era of industrial big data, IEEE Transactions on Industrial Informatics 18 (2) (2021) 976–985.

[10] Alex Kalos, Arthur Kordon, Guido Smits, Sofka Werkmeister, Hybrid model development methodology for industrial soft sensors, in: Proceedings of the 2003 American Control Conference, 2003, vol. 6, IEEE, 2003, pp. 5417–5422.

[11] Qingqiang Sun, Zhiqiang Ge, A survey on deep learning for data-driven soft sensors, IEEE Transactions on Industrial Informatics 17 (9) (2021) 5853–5866.

[12] Weiwu Yan, Di Tang, Yujun Lin, A data-driven soft sensor modeling method based on deep learning and its application, IEEE Transactions on Industrial Electronics 64 (5) (2016) 4237–4245.

[13] Francisco AA Souza, Rui Araújo, Jérôme Mendes, Review of soft sensor methods for regression applications, Chemometrics and Intelligent Laboratory Systems 152 (2016) 69–79.

[14] Zhe Li, Shiyi Qi, Yiduo Li, Zenglin Xu, Revisiting long-term time series forecasting: an investigation on linear mapping, arXiv preprint, arXiv:2305.10721, 2023.

[15] Ailing Zeng, Muxi Chen, Lei Zhang, Qiang Xu, Are transformers effective for time series forecasting?, in: Proceedings of the AAAI Conference on Artificial Intelligence, vol. 37, 2023, pp. 11121–11128.

[16] David Salinas, Valentin Flunkert, Jan Gasthaus, Tim Januschowski, DeepAR: probabilistic forecasting with autoregressive recurrent networks, International Journal of Forecasting 36 (3) (2020) 1181–1191.

[17] Yong Liu, Tengge Hu, Haoran Zhang, Haixu Wu, Shiyu Wang, Lintao Ma, Mingsheng Long, iTransformer: inverted transformers are effective for time series forecasting, arXiv preprint, arXiv:2310.06625, 2023.

[18] Haixu Wu, Jiehui Xu, Jianmin Wang, Mingsheng Long, Autoformer: decomposition transformers with auto-correlation for long-term series forecasting, Advances in Neural Information Processing Systems 34 (2021) 22419–22430.
[19] Yunhao Zhang, Junchi Yan, Crossformer: transformer utilizing cross-dimension dependency for multivariate time series forecasting, in: The Eleventh International Conference on Learning Representations, 2023.
[20] Yuqi Nie, Nam H. Nguyen, Phanwadee Sinthong, Jayant Kalagnanam, A time series is worth 64 words: long-term forecasting with transformers, arXiv preprint, arXiv: 2211.14730, 2022.
[21] Abhimanyu Das, Weihao Kong, Andrew Leach, Shaan Mathur, Rajat Sen, Rose Yu, Long-term forecasting with tide: time-series dense encoder, arXiv preprint, arXiv:2304.08424, 2023.
[22] Haixu Wu, Tengge Hu, Yong Liu, Hang Zhou, Jianmin Wang, Mingsheng Long, TimesNet: temporal 2D-variation modeling for general time series analysis, arXiv preprint, arXiv:2210.02186, 2022.
[23] Tian Zhou, Ziqing Ma, Qingsong Wen, Xue Wang, Liang Sun, Rong Jin, FEDformer: frequency enhanced decomposed transformer for long-term series forecasting, in: International Conference on Machine Learning, PMLR, 2022, pp. 27268–27286.
[24] Zihan Wang, Fanheng Kong, Shi Feng, Ming Wang, Xiaocui Yang, Han Zhao, Daling Wang, Yifei Zhang, Is Mamba effective for time series forecasting?, Neurocomputing 619 (2025) 129178.

CHAPTER 2

Introduction to data-driven prediction

2.1 Overview of single-variable prediction

In the realm of predictive analytics, single-variable prediction stands as a foundational yet crucial approach [1–3]. It involves using a single independent variable to forecast the value of a dependent variable [4]. This method is particularly useful when the relationship between the predictor and outcome is relatively straightforward and can be effectively modeled using a single dimension [5].

Single-variable prediction relies on the assumption that there is a discernible and exploitable relationship between the predictor and target variable [6]. For instance, in aluminum electrolysis, the future cell temperature is linked with the historical temperature values, so we can use the historical observation value to predict the future temperature [7,8]. The housing prediction task, the housing price is linked to the historical price [9–11]. In addition, the stock prediction is also linked to historical stock price [12–14]. The future weather is also linked to the historical observation [15–18]. The aim of single variable prediction is to find the best-fitting linear equation that describes the relationship between the dependent and independent variables. The simplicity of this approach makes it computationally efficient and easy to interpret, which is advantageous for quick decision-making and initial exploratory analysis.

Common techniques used in single-variable prediction include linear regression [19,20], where a straight-line relationship is assumed between the predictor and outcome. It minimizes the sum of the squares of the differences between the observed dependent variable values and those predicted by the linear model [21,22]. These differences are called "residuals". The equation for a simple linear regression model (ordinary least squares regression) can be expressed as

$$y = \beta_0 + \beta_1 x + \epsilon, \tag{2.1}$$

where y is the dependent variable, x is the independent variable, β_0 is the intercept, β_1 is the slope coefficient, and ϵ represents the error term.

Learning-Based Soft Sensing and Predictions for Process Industries
https://doi.org/10.1016/B978-0-44-336759-5.00009-3

2.1.1 Advantages and limitations

The primary advantage of single-variable prediction is its simplicity. It requires minimal data and computational resources, making it accessible even with limited resources. Additionally, the interpretability of the model is high, as the relationship between the predictor and outcome is clear and direct.

However, single-variable prediction also has significant limitations. In many real-world scenarios, the outcome is influenced by multiple factors, not just one. Relying solely on a single variable can lead to inaccurate predictions, as it fails to account for the complexity of the underlying relationship. For example, predicting house prices based only on square footage would ignore other important factors like location, age of the house, and market trends.

Although single-variable prediction serves as a valuable starting point, it often falls short in capturing the full dynamics of complex systems. This limitation naturally leads us to the concept of multi-variable prediction, where multiple independent variables are incorporated to provide a more comprehensive and accurate forecast. By considering additional factors we can better understand the interplay between variables and improve the robustness of our predictive models.

In the following section, we will delve into the intricacies of multi-variable prediction, exploring how it addresses the shortcomings of single-variable prediction and enhances our ability to make informed predictions in diverse contexts.

2.2 Overview of multiple-variable prediction

In the field of predictive analytics, multiple-variable prediction represents a significant advancement over single-variable prediction [23,24]. It involves using multiple independent variables to forecast the value of a dependent variable [25]. This approach is particularly powerful when dealing with complex systems where the outcome is influenced by several factors, rather than just one.

Multiple-variable prediction, also known as multivariate prediction, acknowledges that real-world phenomena are often driven by multiple interrelated factors [24]. By incorporating multiple predictors into a model we can capture the intricate relationships between these variables and the target outcome. For example, in predicting house prices, factors such as square footage, location, age of the house, neighborhood amenities, and market trends can all be considered simultaneously [26,27].

The most common technique for multiple-variable prediction is multiple linear regression [28–30]. The general form of a multiple linear regression model can be expressed as

$$y = \beta_0 + \beta_1 x_1 + \beta_2 x_2 + \cdots + \beta_n x_n + \epsilon, \tag{2.2}$$

where y is the dependent variable, $x_1, x_2, \ldots, x_n$ are the independent variables, β_0 is the intercept, $\beta_1, \beta_2, \ldots, \beta_n$ are the coefficients for each predictor, representing the change in the dependent variable for a one-unit change in the corresponding independent variable, holding other variables constant, and ϵ represents the error term.

The motivation behind the multiple variable prediction is that the sum of the squares of the residual is as small as possible. Mathematically, it is about finding the value of $\beta_1, \beta_2, \ldots, \beta_n$ that minimize this sum.

2.2.1 Advantages and limitations

1. Comprehensive modeling: Multiple-variable prediction allows for a more nuanced understanding of the relationships between various factors and the outcome. By incorporating multiple predictors the model can account for the combined effects and interactions between these variables.
2. Improved accuracy: In many cases, considering multiple variables leads to more accurate predictions. This is particularly true when the dependent variable is influenced by several factors that are not captured by a single predictor.
3. Enhanced interpretability: When the model becomes more complex, the ability to analyze the impact of each independent variable on the outcome provides deeper insights into the underlying dynamics of the system studied.

The limitations of the multiple-variable predictions are d as follows:

1. Complexity and overfitting: As the number of predictors increases, the model becomes more complex. This can lead to overfitting, where the model fits the training data too closely and performs poorly on new, unseen data. Careful model selection and validation techniques are necessary to mitigate this risk.
2. Data requirements: Multiple-variable prediction requires a larger dataset to effectively estimate the coefficients for each predictor.

Insufficient data can lead to unreliable estimates and poor model performance.

3. Multicollinearity: When multiple predictors are highly correlated with each other (multicollinearity), it becomes difficult to isolate the individual effects of each predictor. This can lead to unstable and unreliable coefficient estimates.

Although multiple-variable prediction is a powerful tool, it is not without its challenges. The issue of overfitting, multicollinearity, and the need for large datasets highlight the need for more advanced predictive techniques. In the subsequent sections, we explore these advanced methods, such as machine learning-based algorithms, which help address these limitations and further enhance predictive accuracy.

By building on the foundation of multiple-variable prediction, these advanced techniques allow us to tackle even more complex and high-dimensional problems, ultimately leading to more robust and reliable predictive models.

2.3 Review of time series prediction

Time series prediction (TSP) is a specialized branch of predictive analytics that focuses on forecasting future values based on historical data points collected over time [31–33]. Unlike cross-sectional data, which capture a snapshot at a single point in time, time series data exhibit temporal dependencies and patterns that can be exploited for forecasting purposes.

2.3.1 Challenges and considerations

1. Non-stationarity: Many time series exhibit non-stationary behavior, where the statistical properties change over time. This requires techniques such as differencing or transformation to stabilize the data.
2. Model selection: Choosing the appropriate model and parameters can be complex, especially when dealing with multiple seasonality or non-linear relationships.
3. External factors: Time-series data can be influenced by external events (e.g., economic crises, policy changes) that are not captured by historical data alone.

Addressing these challenges often involves combing time series methods with other predictive techniques, such as machine learning algorithms, to improve forecasting accuracy.

2.4 Review of recursive prediction

Recursive prediction is a method used in forecasting where predictions are made iteratively, using previous predictions as inputs for subsequent forecasts [34,35]. This approach is particularly useful in scenarios where the data are generated sequentially and the prediction horizon extends beyond a single time step. Recursive prediction is commonly applied in fields such as weather forecasting, financial modeling, and system dynamics.

Recursive prediction involves predicting the next time step based on the current and past data and then using this prediction as part of the input for the next iteration. This process is repeated iteratively to generate forecasts for multiple future time steps. For example, in a simple linear recursive prediction model, the forecast for time $t+1$ can be expressed as

$$\hat{y}_{t+1} = f(y_t, \hat{y}_t, y_{t-1}, \hat{y}_{t-1}, \ldots), \tag{2.3}$$

where y_t is the actual value at time t, $\hat{y}_t$ is the predicted value at time t, and f is the prediction function.

2.4.1 Advantages and limitations

The advantages of recursive predictions are:

1. Flexibility: Recursive prediction can handle sequential data and long-term forecasts more naturally than direct prediction methods.
2. Computational efficiency: Recursive models often require less computational power, because they rely on previous predictions rather than solving a large optimization problem at each step.
3. Adaptability: Recursive models can be easily updated as new data become available, making them suitable for real-time forecasting applications.

The limitations of recursive prediction are:

1. Error accumulation: Recursive prediction can accumulate errors over time, especially if the initial predictions are inaccurate. This can lead to significant deviations from the actual values as the forecast horizon increases.
2. Model complexity: Recursive models can become complex, especially when dealing with non-linear relationships and multiple variables. This can make them difficult to interpret and validate.
3. Sensitivity to initial conditions: Recursive prediction is highly sensitive to the initial conditions and the quality of the initial predictions. Small errors in the initial predictions can propagate and magnify over time.

2.4.2 Techniques to improve recursive prediction

There are some key techniques to improve the performance of recursive prediction.

Error correction mechanisms: To mitigate the accumulation of errors, recursive prediction models can incorporate error correction mechanisms. For example, the model can adjust future predictions based on the difference between actual and predicted values at each step.

Hybrid models: Combining recursive prediction with other techniques, such as machine learning algorithms or ensemble methods, can help improve forecasting accuracy. Hybrid models can leverage the strengths of different approaches to produce more robust predictions.

Regularization: Regularization techniques, such as Lasso or Ridge regression, can be applied to recursive prediction models to prevent overfitting and improve generalization performance.

Recursive prediction is a powerful tool for forecasting sequential data and long-term trends. Although it offers several advantages in terms of flexibility and computational efficiency, it also faces challenges related to error accumulation and model complexity. By incorporating error correction mechanisms, combining with other techniques, and applying regularization, we can enhance the performance of recursive prediction models and make more accurate forecasts in various applications.

2.5 Review of direct multiple-horizon prediction

As shown in Fig. 2.1, direct multiple-horizon prediction is a forecasting approach that aims to predict multiple future time steps simultaneously, rather than iteratively as in recursive prediction [36,37]. This method is particularly useful for time series forecasting, where the goal is to generate accurate predictions over a range of future horizons without the potential error accumulation seen in recursive methods [38,39]. In direct multiple-horizon prediction, separate models are trained for each forecast horizon. For example, if the goal is to predict values for the next three time steps, then three distinct models are trained, one for each horizon. This approach avoids the propagation of errors that can occur in recursive prediction, where each subsequent prediction relies on the previous one.

2.5.1 Advantages and limitations

The advantages of multiple-horizon prediction are:

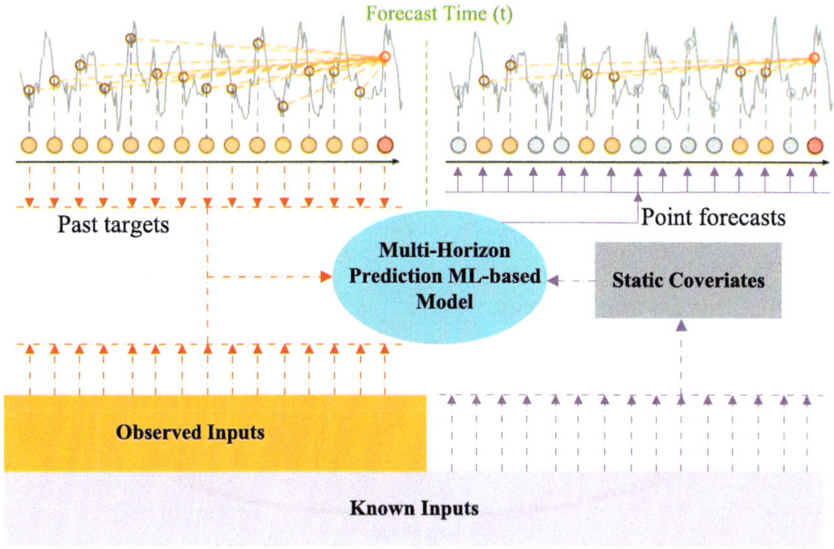

Figure 2.1 Direct multiple-horizon prediction.

1. Error mitigation: By avoiding the iterative use of predictions, direct multiple-horizon prediction reduces the risk of error accumulation.
2. Flexibility: This method can handle different patterns and relationships at each horizon, allowing for more nuanced forecasts.

The limitations are:

1. Computational complexity: Training multiple models can be computationally intensive, especially for long forecast horizons.
2. Data requirements: Each model requires sufficient data to be trained effectively, which can be challenging for short time series.

Direct multiple-horizon prediction offers a powerful alternative to recursive prediction by reducing error accumulation and allowing for more flexible modeling of different horizons. Although it faces challenges related to computational complexity and data requirements, advancements in machine learning and hybrid modeling techniques continue to enhance its applicability and effectiveness in time series forecasting.

2.5.2 Recursive vs. direct prediction

Fig. 2.2 describes the recursive and direct prediction strategies. Recursive prediction differs from direct prediction, where the model directly predicts the value at a future time step without relying on intermediate predictions.

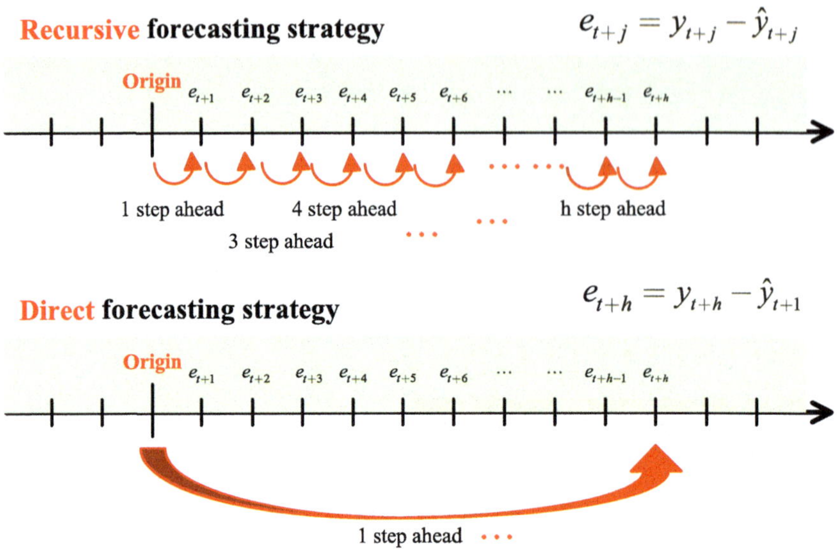

Figure 2.2 Recursive vs direct forecasting strategies.

Recursive prediction has the advantage of being more flexible and computationally efficient, especially for long-term forecasts. However, it can also accumulate errors over time, leading to less accurate predictions as the forecast horizon increases.

2.6 Conclusion

This chapter introduces five distinct data-driven prediction methods, providing the readers with their definitions and notation. The following chapter presents two typical industrial cases and demonstrates how to apply these prediction methods in such scenarios.

References

[1] Jia Wang, Vincent W. Zheng, Zemin Liu, Kevin Chen-Chuan Chang, Topological recurrent neural network for diffusion prediction, in: 2017 IEEE International Conference on Data Mining (ICDM), IEEE, 2017, pp. 475–484.

[2] Georg Börner, Hauke Haehne, Jose Casadiego, Marc Timme, Revealing system dimension from single-variable time series, Chaos: An Interdisciplinary Journal of Nonlinear Science 33 (7) (2023).

[3] José F. Torres, Dalil Hadjout, Abderrazak Sebaa, Francisco Martínez-Álvarez, Alicia Troncoso, Deep learning for time series forecasting: a survey, Big Data 9 (1) (2021)

3–21.
[4] Ahmad Alsharef, Karan Aggarwal, Sonia, Manoj Kumar, Ashutosh Mishra, Review of ML and AutoML solutions to forecast time-series data, Archives of Computational Methods in Engineering 29 (7) (2022) 5297–5311.
[5] Zijian Zhang, Koustav Rudra, Avishek Anand, Explain and predict, and then predict again, in: Proceedings of the 14th ACM International Conference on Web Search and Data Mining, 2021, pp. 418–426.
[6] Han Ding, Zeli Li, Qiuru Ren, Haitao Chen, Menglai Song, Yuqiu Wang, Single-variable method for predicting trends in chlorophyll a concentration based on the similarity of time series, Ecological Indicators 140 (2022) 109027.
[7] Yongxiang Lei, Xiaofang Chen, Ming Min, et al., A semi-supervised Laplacian extreme learning machine and feature fusion with CNN for industrial superheat identification, Neurocomputing 381 (2020) 186–195.
[8] Yongxiang Lei, Hamid Reza Karimi, Lihui Cen, Xiaofang Chen, Yongfang Xie, Processes soft modeling based on stacked autoencoders and wavelet extreme learning machine for aluminum plant-wide application, Control Engineering Practice 108 (2021) 104706.
[9] P. Vidya Sree, Predictive modeling for housing price trends using historical data, International Scientific Journal of Engineering and Management (2024), https://doi.org/10.55041/isjem01418.
[10] Prof. Anup Bhange, Rajvaibhav Patil, M.D. Masoom Khan, Prajwal Pazare, Harsh Patil, House price prediction, International Journal for Research in Applied Science and Engineering Technology (2024), https://doi.org/10.22214/ijraset.2024.66042.
[11] Prof. N.R. Jain, Prachi S. Zende, Mohini Gaikwad, Vibha Bhor, Housing price prediction using machine learning. 2020.
[12] Yumo Xu, Shay B. Cohen, Stock movement prediction from tweets and historical prices, 2018, pp. 1970–1979.
[13] Huajin Liu, Tianlong Zhao, Suwei Wang, Xuemei Li, A stock rank prediction method combining industry attributes and price data of stocks, Information Processing & Management 60 (2023) 103358, https://doi.org/10.1016/j.ipm.2023.103358.
[14] Chuanjun Zhao, Meili Wu, Jingfeng Liu, Zening Duan, Jie Li, Lihua Shen, Xuekui Shangguan, Donghang Liu, Yanjie Wang, Progress and prospects of data-driven stock price forecasting research, International Journal of Cognitive Computing in Engineering (2023), https://doi.org/10.1016/j.ijcce.2023.03.001.
[15] A. Ribes, S. Qasmi, N. Gillett, Making climate projections conditional on historical observations, Science Advances 7 (2021), https://doi.org/10.1126/sciadv.abc0671.
[16] Andrew D. Jones, Deeksha Rastogi, P. Vahmani, A. Stansfield, Kevin A. Reed, Travis B. Thurber, Paul A. Ullrich, J. Rice, Continental United States climate projections based on thermodynamic modification of historical weather, Scientific Data 10 (2023), https://doi.org/10.1038/s41597-023-02485-5.
[17] Anthony McNally, Christian Lessig, Peter Lean, Eulalie Boucher, Mihai Alexe, E. Pinnington, M. Chantry, Simon Lang, Chris Burrows, Marcin Chrust, F. Pinault, Ethel Villeneuve, Niels Bormann, S. Healy, Data driven weather forecasts trained and initialised directly from observations, arXiv, arXiv:2407.15586 [abs], 2024, https://doi.org/10.48550/arXiv.2407.15586.
[18] Eugénio Rodrigues, Marco S. Fernandes, D. Carvalho, Future weather generator for building performance research: an open-source morphing tool and an application, in: Building and Environment, 2023.
[19] Gareth James, Daniela Witten, Trevor Hastie, Robert Tibshirani, Jonathan Taylor, Linear regression, in: An Introduction to Statistical Learning: With Applications in Python, Springer, 2023, pp. 69–134.
[20] Douglas C. Montgomery, Elizabeth A. Peck, G. Geoffrey Vining, Introduction to Linear Regression Analysis, John Wiley & Sons, 2021.

[21] Seyedeh Reyhaneh Shams, Ali Jahani, Saba Kalantary, Mazaher Moeinaddini, Nematollah Khorasani, The evaluation on artificial neural networks (ANN) and multiple linear regressions (MLR) models for predicting SO_2 concentration, Urban Climate 37 (2021) 100837.

[22] Paul Goldsmith-Pinkham, Peter Hull, Michal Kolesár, Contamination bias in linear regressions, American Economic Review 114 (12) (2024) 4015–4051.

[23] Martin Krzywinski, Naomi Altman, Multiple linear regression: when multiple variables are associated with a response, the interpretation of a prediction equation is seldom simple, Nature Methods 12 (12) (2015) 1103–1105.

[24] Jason W. Osborne, Prediction in multiple regression, Practical Assessment, Research & Evaluation 7 (1) (2000).

[25] Leo Breiman, Jerome H. Friedman, Predicting multivariate responses in multiple linear regression, Journal of the Royal Statistical Society, Series B, Statistical Methodology 59 (1) (1997) 3–54.

[26] Qingqi Zhang, Housing price prediction based on multiple linear regression, Scientific Programming 2021 (1) (2021) 7678931.

[27] C.H. Raga Madhuri, G. Anuradha, M. Vani Pujitha, House price prediction using regression techniques: a comparative study, in: 2019 International Conference on Smart Structures and Systems (ICSSS), IEEE, 2019, pp. 1–5.

[28] Mark Tranmer, Mark Elliot, Multiple linear regression, The Cathie Marsh Centre for Census and Survey Research (CCSR) 5 (5) (2008) 1–5.

[29] Lynn E. Eberly, Multiple linear regression, Topics in Biostatistics (2007) 165–187.

[30] Gülden Kaya Uyanık, Neşe Güler, A study on multiple linear regression analysis, Procedia – Social and Behavioral Sciences 106 (2013) 234–240.

[31] Zhongyang Han, Jun Zhao, Henry Leung, King Fai Ma, Wei Wang, A review of deep learning models for time series prediction, IEEE Sensors Journal 21 (6) (2019) 7833–7848.

[32] Muhammad Faheem Mushtaq, Urooj Akram, Muhammad Aamir, Haseeb Ali, Muhammad Zulqarnain, Neural network techniques for time series prediction: a review, JOIV: International Journal on Informatics Visualization 3 (3) (2019) 314–320.

[33] Spyros Makridakis, Time series prediction: forecasting the future and understanding the past, International Journal of Forecasting 10 (3) (1994) 463–466.

[34] Luis Javier Herrera, Hector Pomares, Ignacio Rojas, Alberto Guillén, Alberto Prieto, Olga Valenzuela, Recursive prediction for long term time series forecasting using advanced models, Neurocomputing 70 (16–18) (2007) 2870–2880.

[35] Andreas Zelios, Achilleas Grammenos, Maria Papatsimouli, Nikolaos Asimopoulos, George Fragulis, Recursive neural networks: recent results and applications, in: SHS Web of Conferences, vol. 139, EDP Sciences, 2022, p. 03007.

[36] Ke-Li Xu, Testing for multiple-horizon predictability: direct regression based versus implication based, The Review of Financial Studies 33 (9) (2020) 4403–4443.

[37] Yongxiang Lei, Hamid Reza Karimi, Underflow concentration soft sensing for cone thickener system based on a direct data-driven quantile regression forecasting, Mechatronics Technology 2 (1) (2025) 19–20.

[38] Jiawen Zhang, Shun Zheng, Xumeng Wen, Xiaofang Zhou, Jiang Bian, Jia Li, ElasTST: towards robust varied-horizon forecasting with elastic time-series transformer, Advances in Neural Information Processing Systems 37 (2024) 119174–119197.

[39] Thangjam Aditya, Sanjita Jaipuria, Pradeep Kumar Dadabada, A review of methods for long-term electric load forecasting, Journal of Forecasting (2024).

CHAPTER 3

Aluminum electrolysis industry

3.1 Theory of aluminum electrolysis

The aluminum industry is a strategic pillar of national development, military capability, aerospace advancement, and agricultural infrastructure [1,2]. Aluminum alloys have been the focus of extensive investigation for many decades, owing to their exceptional strength-to-weight ratio, corrosion resistance, fatigue tolerance, recyclability, and ease of fabrication features that underpin their widespread application across aircraft, defense systems, infrastructure, and sustainable agriculture [3,4].

The aluminum production workshop is presented in Fig. 3.1. In aluminum electrolysis process the production variables (or key quality variables) are linked to each other [5,6]. The coupled variables often existed in a harsh environment, such as high temperature, high magnificence, strong coupled and large delay, and complex environments. In the production of industrial aluminum the degree of superheat (SD) is the difference between the cell temperature and the crystal situation [7–9]. It is a very important indicator for monitoring the whole cell situation, improving the quality of aluminum, and leveraging the electronic power cost. However, some key parameters cannot be directly measured because of the influence of a harsh environment, high temperature, and strong magnets. In practical aluminum production the crystal temperature requires an offline laboratory analysis to get an accurate value, improving the

Figure 3.1 The aluminum electrolysis reduction workshop.

Learning-Based Soft Sensing and Predictions for Process Industries
https://doi.org/10.1016/B978-0-44-336759-5.00010-X

detection cost of SD [10]. Knowing the reaction mechanism of the aluminum process, we can see that there is some inner link between these variables and the output key variable, which gives us the insight to develop some novel learning-based frameworks for soft modeling in the industrial aluminum plant.

Fig. 3.2 is the process flow diagram of the aluminum electrolysis process. The main reactions in the bath can be given as

$$2Al_2O_3 + 3C = 4Al + 3CO_2, \tag{3.1}$$

$$Al_2O_3 + 3C = 2Al + 3CO. \tag{3.2}$$

We see that the alumina (Al_2O_3) and the carbon anode are the main reaction raw materials in the aluminum reduction cell [11]. However, if the alumina power has a high liquidus temperature (up to 1011°C), a high electronic energy will be wasted. To lower the energy consumption, aluminum fluoride (AlF_3) and calcium fluoride (CaF_2) are added to the bath, and hence some subsidiary reactions are

$$3CO_3^{2-} + 2AlF_6^{3-} \rightarrow 12F^- + Al_2O_3 \downarrow + 3CO_2 \uparrow, \tag{3.3}$$

$$3CO_3^{2-} + 2AlF_3 \rightarrow 6F^- + Al_2O_3 \downarrow + 3CO_2 \uparrow, \tag{3.4}$$

and the total neutralization reaction is

$$3Li_2^{2+} + 2AlF_6^{3-} \rightarrow 6F^- + 6LiF \downarrow + Al_2O_3 \downarrow + 3CO_2 \uparrow. \tag{3.5}$$

All these complex reaction environments make it hard to directly detect the SD situation. SD detection very often needs two steps. The first stage requires a manual operation by using the thermal device to get the bath temperature, and the second stage is an off-line laboratory analysis of the crystal temperature. Both steps are economic and time-consuming. SD situation is classified into three categories, high label, middle label, and low label. A high label means that the difference between pot and crystal temperatures is greater than 15°C. The middle label in the SD range is 8–12°C, and the low label means that the difference is lower than 8°C.

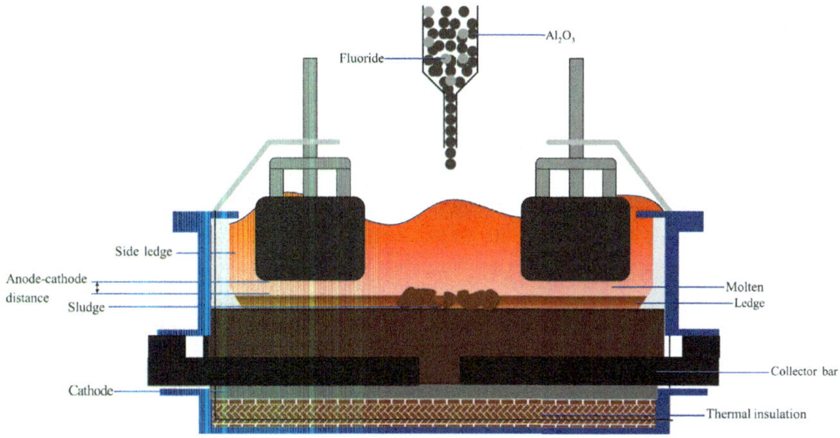

Figure 3.2 Flowchart diagram of the aluminum production unit.

3.2 Prediction requirements and challenges

3.2.1 Description of prediction tasks

In a time-series prediction task, a learning model seeks to capture dependencies among input variables. Some of these variables are directly observed, denoted $\{x_1, \ldots, x_N\}$, whereas others are hidden or only intermittently observed, denoted $\{x_{N+1}, \ldots, x_{N+s}\}$. Our objective is to design a model or framework that, based on historical data (both labeled and unlabeled), detects or forecasts a target quality variable Y.

In the aluminum electrolysis process, key variables evolve dynamically in a complex electrochemical environment, yet they remain interdependent (see Table 3.1). This interdependence enables data-driven models to learn functional mappings from observed inputs to the target quality variable, such as superheat degree (SD) or bath temperature.

Given a multivariate time-series $\mathbf{x}_t = \{x_{1,t}, \ldots, x_{N,t}\}$ with partial observations of hidden variables, the task is to estimate or predict the target

$$Y_t = f(\mathbf{x}_{t-T:t}, \boldsymbol{\theta}),$$

where T is the historical window length, and $\boldsymbol{\theta}$ are the model parameters to be learned from data.

Commonly used architectures in such tasks include:

- **Recurrent neural networks (RNN/LSTM/GRU)**: Capture temporal dependencies in sequence data, for example, apply deep LSTM models to predict bath temperature in aluminum electrolysis cells [11,12].

- **Temporal convolutional networks (TCN)**: Offer improved parallelism and stability compared to RNNs and are adapted for real-time bath temperature estimation [13].
- **Hybrid attention mechanisms**: Dual-attention LSTM combined with ARMA effectively handles non-stationary subsequences in cell temperature time series [14].
- **A dual-attention LSTM and ARMA hybrid model** to predict aluminum cell temperature, achieving less than five-degree RMSE, which is suitable for real-time control [14].
- **The self-supervised dilated LSTM (DDLSTM) model** proposed by Lei et al. [12] for electrolyte temperature prediction; by exploiting unlabeled data, it improves robustness in SD or temperature estimation tasks.
- **Sparse neural nets with ℓ_1 regularization and skip-connections** model electrolysis dynamics with better interpretability and reliable long-term forecasting [15,16].

We give a workflow overview.

1. *Data preprocessing:* Resample and normalize time-series measurements, handle missing values, and optionally cluster cells by operating conditions.
2. *Feature engineering:* Use lagged variables, rolling statistics, and optionally auto-extracted features via CNNs.
3. *Model selection:* Choose architectures (LSTM, TCN, hybrid attention) based on dependency structures, label availability, and real-time demands.
4. *Training and validation:* Employ train/validate splits or *k*-fold cross-validation, using metrics like RMSE, MAE, and, if classification-based, precision/recall.
5. *Deployment:* Integrate the trained model into a DCS or SCADA system for live inference, enabling proactive control adjustments (e.g., feed rate or current regulation).

Accurate prediction of SD or bath temperature—based on observed variables—allows:

- Proactive correction of thermal deviations.
- Energy efficiency improvements by avoiding overcompensation.
- Reduced reliance on coarse intermittent laboratory sampling.

Table 3.1 Process variables selected based on operator experience.

No.	Tags	Descriptions
1	u_1	Working current
2	u_2	Cell resistance
3	u_3	Set datum time of Al_2O_3
4	u_4	Actual feeding time of Al_2O_3
5	u_5	Content of iron
6	u_6	Content of silica
7	u_7	Aluminum tapping level
8	u_8	Electrolyte level
9	u_9	Content of precipitated aluminum
10	u_{10}	Molecular radio
11	u_{11}	Average cell voltage
12	u_{12}	Addition amount of AlF_3

For example, hybrid attention models and sparse neural nets have demonstrated RMSE below 5 degrees in cell temperature prediction [14], and Lei's DDLSTM yields strong performance despite scarce labels [12].

Time-series prediction in aluminum electrolysis—especially for variables like SD and temperature—poses unique challenges: complex dynamics, unobserved states, and limited labels. Carefully designed architectures (LSTM, TCN, sparse networks) and workflows can deliver robust forecasting and enhanced process control.

3.2.2 Description of soft sensing tasks

In the aluminum electrolysis reduction, there key quality variables that are hard to measure by the physical devices monitored by the data-driven "soft sensing" techniques. The superheat degree, one of the suitable soft sensing variables, is linked to historical bath temperature and other key variables and therefore can be modeled in a soft sensing task.

In the Hall–Héroult aluminum electrolysis process, many important quality variables—such as electrolyte temperature, aluminum fluoride concentration, bath level, and especially the *superheat degree* (SD)—are difficult or impractical to measure continuously with physical sensors [17,18]. The harsh environment inside an electrolysis cell—high temperature, strong magnetic fields, and corrosive electrolyte—rapidly destroys thermocouples and other devices, so these variables are typically measured intermittently or estimated offline [19–21].

A *soft sensor*, or data-driven estimator, uses easily available online measurements (e.g., bath temperature, cell voltage, current, flame-hole video frames) to infer such hidden process variables within a data-driven modeling framework [10]. Among them, SD, defined as the difference between bath temperature and the crystallization (liquidus) temperature of the molten electrolyte, is a critical indicator. It directly reflects the cell's thermal balance and current efficiency, as too low SD leads to freezing of cryolite and high anode effect rates, whereas too high SD degrades current efficiency and aluminum quality [22,23].

Real-time monitoring of SD is essential for optimizing energy consumption and ensuring stable operation [9]. However, since the crystallization temperature cannot be measured directly, direct online SD measurement is infeasible. Instead, experienced operators visually inspect the flame-hole via cameras and infer SD from color and fluctuation patterns, a method that is subjective, non-standard, and prone to inconsistencies [24]. This motivates development of ML-based soft sensors for SD.

A variety of data-driven architectures have been proposed:

- CNN-based soft sensors using flame-hole video, often combined with extreme learning machine (ELM) or kernel ELM. For instance, the CNN–ELM framework achieved classification accuracies above 85% on SD states [25,22].
- Semi-supervised Laplacian ELM methods extract spatiotemporal video features with CNN, followed by manifold regularization to improve generalization, reaching ~89% accuracy [25,10].
- Self-supervised LSTM: models historical temperature and other process time series to forecast SD, especially under intermittent renewable-powered operation [12,13].
- Digital Twin + NMPC: physics-informed or hybrid models estimate electrolyte temperature and hence SD in real-time using thermocouples embedded in pot linings combined with model-predictive estimation [26].

The development workflow is commonly shown as follows:

1. **Variable selection:** real-time sensors (bath temperature, cell voltage, current, electrolyte additions) and video data.
2. **Feature extraction:** 2D/3D CNNs for video; time-series lags and temporal features for process data.

3. **Modeling:** ELM, kernel ELM, SVM, ANN, LSTM, or CNN–LSTM hybrids. Semi-supervised extensions improve performance when labels are scarce [12].
4. **Training and validation:** using labeled historical data with offline SD measurements or expert annotations and evaluating via RMSE, MAE, and classification accuracy.
5. **Deployment:** integrating trained models into DCS for real-time inference, enabling closed-loop optimization and energy-efficient control.

Soft sensing of superheat degree in aluminum electrolysis is a practical problem with clear industrial significance [27,28]. By leveraging modern data-driven approaches—particularly CNN and LSTM-based models—the industry can achieve accurate and robust online estimations of SD. This capability opens the door for improved process control, reduced manual intervention, enhanced energy efficiency, and safer, more consistent operation [29].

3.3 Conclusions

This chapter has delved into the aluminum electrolysis industry, introducing the theoretical foundations of the aluminum production process. Key challenges inherent to the industry were also outlined. Finally, the critical tasks of prediction and soft sensing within this production environment were presented. Addressing these prediction and soft sensing challenges is essential for optimizing aluminum electrolysis operations. The next chapter shifts focus to another industrial process, the mining paste-filling process.

References

[1] Xiaodong Wu, Wenkang Zhang, A review on aluminum matrix composites' characteristics and applications for automotive sector, Heliyon 10 (20) (2024).

[2] Nikunj Patel, Gaurav Mittal, Mitushi Agrawal, Ajaya Kumar Pradhan, Aluminum foam production, properties, and applications: a review, International Journal of Metalcasting 18 (3) (2024) 2181–2198.

[3] Xin Li, Jingkai Lin, Chang Liu, Aimin Liu, Zhongning Shi, Zhaowen Wang, Shaoyan Jiang, Gang Wang, Fengguo Liu, Research on aluminum electrolysis from 1970 to 2023: a bibliometric analysis, JOM 76 (7) (2024) 3265–3274.

[4] Edgar A. Starke Jr, J.T. Staley, Application of modern aluminum alloys to aircraft, Progress in Aerospace Sciences 32 (2–3) (1996) 131–172.

[5] Yongxiang Lei, Hamid Reza Karimi, Lihui Cen, Xiaofang Chen, Yongfang Xie, Processes soft modeling based on stacked autoencoders and wavelet extreme learning

machine for aluminum plant-wide application, Control Engineering Practice 108 (2021) 104706.
[6] Xu Yang, Yue Zhang, Yuri A.W. Shardt, Xiaoli Li, Jiarui Cui, Chaonan Tong, A KPI-based soft sensor development approach incorporating infrequent, variable time delayed measurements, IEEE Transactions on Control Systems Technology 28 (6) (2019) 2523–2531.
[7] Weichao Yue, Weihua Gui, Xiaofang Chen, Zhaohui Zeng, Yongfang Xie, Evaluation strategy and mass balance for making decision about the amount of aluminum fluoride addition based on superheat degree, Journal of Industrial and Management Optimization 16 (2) (2020) 601.
[8] Weichao Yue, Lingfeng Hou, Xiaoxue Wan, Xiaofang Chen, Weihua Gui, Superheat degree recognition of aluminum electrolysis cell using unbalance double hierarchy hesitant linguistic Petri nets, IEEE Transactions on Instrumentation and Measurement 72 (2023) 1–15.
[9] J. Wang, S. Xie, Y. Xie, et al., Label propagation with contrastive anchors for deep semi-supervised superheat degree identification in aluminum electrolysis process, IEEE Transactions on Automation Science and Engineering 21 (2) (2023) 1284–1296, https://doi.org/10.1109/TASE.2023.3241234.
[10] Yongxiang Lei, Xiaofang Chen, Yongfang Xie, An improved cell situation identification approach with convolutional neural network and wavelet extreme learning machine, in: Proceedings of the Institution of Mechanical Engineers, Part i: Journal of Systems and Control Engineering, 2020, 0959651820935667.
[11] Svein Kolås, Tor Støre, Bath temperature and AlF_3 control of an aluminium electrolysis cell, Control Engineering Practice 17 (9) (2009) 1035–1043, https://doi.org/10.1016/j.conengprac.2009.04.002.
[12] Yongxiang Lei, Hamid Reza Karimi, Xiaofang Chen, A novel self-supervised deep LSTM network for industrial temperature prediction in aluminum processes application, Neurocomputing 502 (2022) 177–185.
[13] Yongxiang Lei, Xiaofang Chen, Yongfang Xie, Lihui Cen, A self-supervised temporal temperature prediction method based on dilated contrastive learning, Journal of Process Control 120 (2022) 150–158.
[14] Ye Zhu, Shiwen Xie, Yongfang Xie, Xiaofang Chen, Temperature prediction of aluminum reduction cell based on integration of dual attention LSTM for non-stationary sub-sequence and ARMA for stationary sub-sequences, Control Engineering Practice 138 (2023) 105567.
[15] Changqing Yuan, Yongfang Xie, Shiwen Xie, Jie Wang, Pruned tree-structured temporal convolutional networks for quality variable prediction of industrial process, Journal of Process Control (2024), https://doi.org/10.1016/j.jprocont.2024.103312.
[16] Zuguo Chen, Yong Gang Li, Xiaofang Chen, Chunhua Yang, W. Gui, Anode effect prediction based on collaborative two-dimensional forecast model in aluminum electrolysis production, Journal of Industrial and Management Optimization (2019), https://doi.org/10.3934/jimo.2018060.
[17] L.J.D.S. Moreira, M. Fiacchini, G. Besançon, Francesco Ferrante, H. Roustan, State affine modeling and observer design for Hall-Héroult process, IFAC-PapersOnLine 53 (2020) 12020–12025, https://doi.org/10.1016/J.IFACOL.2020.12.736.
[18] L.J.D.S. Moreira, G. Besançon, Francesco Ferrante, M. Fiacchini, H. Roustan, Convection-diffusion model for alumina concentration in Hall-Héroult process★, IFAC-PapersOnLine (2022), https://doi.org/10.1016/j.ifacol.2022.09.259.
[19] Jie Wang, Yongfang Xie, Shiwen Xie, Xiaofang Chen, Development of data-knowledge-driven predictive model and multi-objective optimization for intelligent optimal control of aluminum electrolysis process, Engineering Applications of Artificial Intelligence 134 (2024) 108664, https://doi.org/10.1016/j.engappai.2024.108664.

[20] E. Lundby, A. Rasheed, J. Gravdahl, I. Halvorsen, A novel hybrid analysis and modeling approach applied to aluminum electrolysis process, Journal of Process Control 105 (2021) 62–77, https://doi.org/10.1016/J.JPROCONT.2021.06.005.

[21] Haakon Robinson, E. Lundby, A. Rasheed, J. Gravdahl, Deep learning assisted physics-based modeling of aluminum extraction process, Engineering Applications of Artificial Intelligence 125 (2023) 106623, https://doi.org/10.1016/j.engappai.2023.106623.

[22] Yongxiang Lei, Xiaofang Chen, Ming Min, et al., A semi-supervised Laplacian extreme learning machine and feature fusion with cnn for industrial superheat identification, Neurocomputing 381 (2020) 186–195.

[23] Yongxiang Lei, Xiaofang Chen, Yongfang Xie, An improved cell situation identification approach with convolutional neural network and wavelet extreme learning machine, Proceedings of the Institution of Mechanical Engineers. Part I, Journal of Systems and Control Engineering 235 (10) (2021) 1898–1905.

[24] Shiwen Xie, Yongfang Xie, Tingwen Huang, Identification of superheat degree for aluminum electrolysis based on feature-guide deep variational autoencoder, in: 2024 6th International Conference on Industrial Artificial Intelligence (IAI), 2024, pp. 1–6.

[25] Weichao Yue, Haiyang Hu, Xiaoxue Wan, Xiaofang Chen, Weihua Gui, A domain knowledge-supervised framework based on deep probabilistic generation network for enhancing industrial soft-sensing, IEEE Transactions on Instrumentation and Measurement (2025).

[26] Ye Zhu, Shiwen Xie, Yongfang Xie, Xiaofang Chen, Recognition of aluminum electrolysis overheat trend based on DA-LSTM neural network, in: 2021 8th International Conference on Information, Cybernetics, and Computational Social Systems (ICCSS), 2021, pp. 290–295.

[27] Xiaojun Li, Bo Liu, Wenbo Qian, et al., Design of soft-sensing model for alumina concentration based on improved deep belief network, Processes 10 (12) (2022) 2537, https://doi.org/10.3390/pr10122537.

[28] H.S. Li, C.W. Jiang, Development and application of soft sensor model for heterogeneous information of aluminum reduction cells, Control Engineering Practice 19 (10) (2011) 1109–1115, https://doi.org/10.1016/j.conengprac.2011.05.002.

[29] Yiyin Tang, Yalin Wang, Chenliang Liu, Xiaofeng Yuan, Kai Wang, Chunhua Yang, Semi-supervised LSTM with historical feature fusion attention for temporal sequence dynamic modeling in industrial processes, Engineering Applications of Artificial Intelligence 117 (2023) 105547.

CHAPTER 4

Paste-filling industry

4.1 Theory of paste-filling system

The paste-filling system is a critical component in modern mining operations, playing an essential role in tailings management and underground backfilling. Fig. 4.1 illustrates a laboratory-scale deep cone thickener system. As shown in Fig. 4.2, there is a complex and coupled nonlinear reaction in this device. The deep CTS is crucial for subsurface paste filling to ensure consistent concentration levels for underground mining fills [1–3]. Maintaining the appropriate subterranean concentration is essential, as pipe blockages can occur when the concentration is too high, whereas low concentration reduces the quality of the backfill paste, posing safety hazards for mining operations [4,5]. Therefore the priority is developing a model to predict underflow concentrations in the CTS system [6]. The CTS operates on a continuous and hierarchical concept, with a crude low-concentration slurry flow (around 20%–30%) fed into it, combined with a flocculant to speed up the sinking rate and collect dissolved particles in a mud bed [7]. The bottom of the CTS generates a suitable concentration and volume feed flow, whereas the top clean water from the overflow pipe is recycled for future use [8]. The primary control requirement for the CTS is maintaining a consistent and precise underflow concentration, which is critical to assessing the effectiveness and efficiency of the underground pasting process [9]. To anticipate the underflow concentration, it is necessary to identify changes in various variables due to the inner link of the production quality variable in the deep cone thickening process [10,11]. A time-series architecture can be used to obtain prior knowledge and historical information, allowing for accurate underflow concentration prediction. All the variables in this process are listed in Table 4.1.

4.2 Prediction requirements and challenges

4.2.1 Description of prediction tasks

To reconcile and improve upon these separate methods, a new sequence-to-sequence direct data-driven framework that generates multiple horizon

Learning-Based Soft Sensing and Predictions for Process Industries
https://doi.org/10.1016/B978-0-44-336759-5.00011-1

Figure 4.1 Laboratory deep cone thickener.

quantile forecasts with uncertainty consideration for the industrial DCS is presented [1]. It is designed to solve the large-scale time series regression problem

$$p(y_{t+k,i}, \ldots, y_{t+1,i} \mid y_{:t,i}, x^{(h)}_{:t,i}, x^{(f)}_{t:,i}, x^{(s)}_{i}), \tag{4.1}$$

where $y_{\cdot,i}$ is the ith time series to forecast, $x^{(h)}_{:t,i}$ are the temporal covariates available in history, $x^{(f)}_{t:,i}$ is the knowledge about the future, and $x^{(s)}_{i}$ are the static, time-invariant features. Each series is considered as one sample fed into a single time sequence neural network, such as a recurrent neural network (RNN) or convolutional neural network (CNN), even if they correspond to different items. This enables cross-series learning and cold-start forecasting for items with limited history.

In the paste-filling process the underflow concentration is coupled with many other variables [12,13]. In practice, there are numerous interdependent variables, and their precise relationship with the target variable is often unknown, making it challenging to determine which variables are relevant for prediction. It is also difficult to determine the

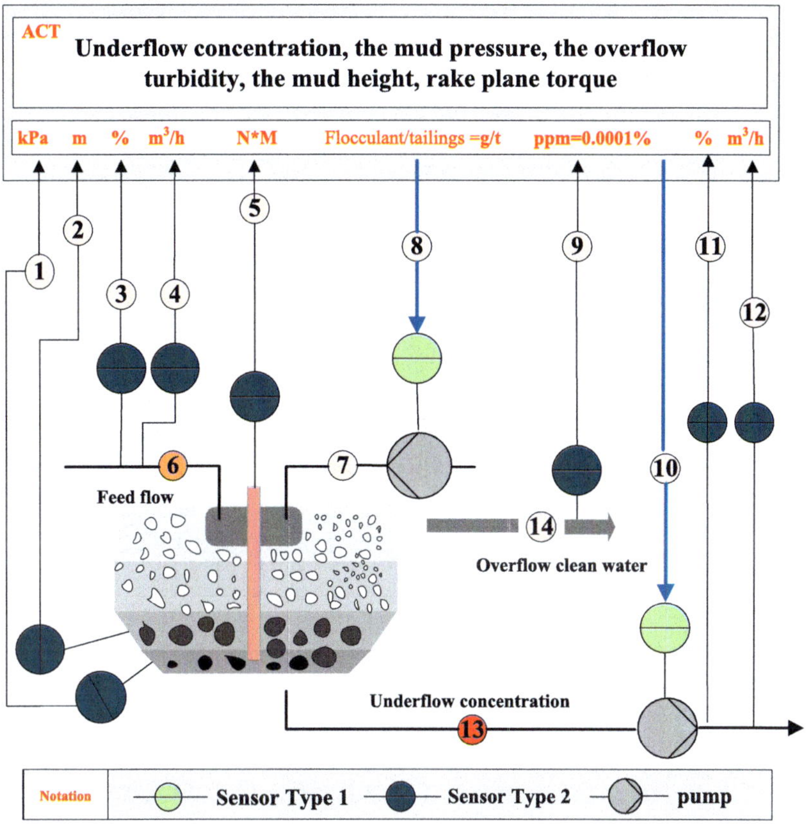

Figure 4.2 Flowchart of the deep cone thickener production.

necessary prediction horizon and historical observation. To address this, a coefficient analysis was conducted to identify the essential variables in the cone thickener production process. The selection of auxiliary variables is a critical first step in the development of a data-driven prediction model. The mud height and pressure are important factors added to the model due to their non-linear relationship, as determined by mechanical analysis. The final selection of the dominant variables includes the mud pressure, mud height, last instant feed flow, maximum volume flow, underflow density, and front instant underflow concentration. Developed based on [12], the input variables are the mud pressure, the mud height, the last instant feed flow, the flow of the highest volume, the underflow density, and the historical underflow concentration. Therefore the underflow concentration multi-horizon prediction in CTS can be given as follows:

Table 4.1 Variable candidates specification.

Symbol	Quality	Description
①	Mud pressure	KPa
②	Mud height level	m
③	Feed flow concentration	$C\%$
④	Feed flow amount	Q
⑤	The rake torque	$N \cdot M$
⑥	Tailings	t
⑦	The amount of flocculant	g/t
⑧	Flocculant flowrate	m/s
⑨	Overflow turbidity	ppm
⑩	The rotation speed of rake	r/min
⑪	Frontier *t*th underflow concentration	$C\%$
⑫	Underflow flow amount	Q
⑬	Underflow pressure	KPa
⑭	Overflow channel	$C\%$

$$\begin{aligned} & p(y_{t+k,i}, \ldots, y_{t+1,i} \mid y_{:t,i}, x_{:t,i}^{(h)}, x_{t:,i}^{(f)}, x_i^{(s)}) \\ & = p(y_{t+k,i}, \cdots, y_{t+1,i} \mid \langle \mathbf{C}_{i(t)}, \mathbf{Q}_{i(t)}, \mathbf{F}_{(t)}, \mathbf{C}_{o(t)}, \mathbf{Q}_{o(t)} \rangle) \\ & = p(y_{t+k,i}, \ldots, y_{t+1,i} \mid \mathbf{W} \cdot NN \sum_{i=1}^{L} \langle y_i, y_i' \rangle), \end{aligned} \tag{4.2}$$

where NN represents the neural network-based predictor, and $\mathbf{W}$ is the corresponding weight vector. The input variables are denoted by the matrix $[C_i, Q_i, F, C_o, Q_o]$, which denotes the input flow concentration, input flow, mud pressure, underflow concentration, and underflow amount, respectively. However, in many scenarios, the provision for prediction intervals is very insightful for the control and operation, see Table 4.1. So Eq. (4.2) with quantile addition can be rewritten as

$$\begin{aligned} & p(y_{t+k,i}, \ldots, y_{t+1,i} \mid y_{:t,i}, x_{:t,i}^{(h)}, x_{t:,i}^{(f)}, x_i^{(s)}, q) \\ & = p(y_{t+k,i}, \ldots, y_{t+1,i} \mid \langle \mathbf{C}_{i(t)}, \mathbf{Q}_{i(t)}, \mathbf{F}_{(t)}, \mathbf{C}_{o(t)}, \mathbf{Q}_{o(t)} \rangle, q) \\ & = p(y_{t+k,i}, \ldots, y_{t+1,i} \mid \mathbf{W} \cdot NN \sum_{i=1}^{L} \langle y_i, y_i' \rangle, \end{aligned} \tag{4.3}$$

which implements the predicted qth quantile of the k-step-ahead forecast at instant t.

4.2.2 Description of soft sensing tasks

In underground mining operations, paste-filling is a critical process for achieving safe and efficient ore extraction and ground control. This process involves mixing mill tailings, cementitious materials (typically cement), and water to prepare a paste with specific rheological properties, which is then pumped through pipelines into mined-out stopes. The deep cone thickener (DCT), serving as the key front-end equipment for paste preparation, is responsible for efficiently concentrating low-density tailings slurry to the target high density. Its operational state directly determines paste quality and subsequent pumping performance.

The operation of a deep cone thickener is a highly complex, multivariable coupled dynamic process. It encompasses multiple physicochemical sub-processes internally, including solid–liquid separation, particle settling, compression, and clarification. To achieve precise control and optimization of thickener performance, thereby ensuring the quality of the final paste (e.g., density, stability, pumpability), it is essential to monitor the real-time status of several internal key quality variables (KQVs). However, direct, continuous, and reliable online measurement of these critical variables faces significant challenges:

- **Underflow concentration:** This is the thickener's most crucial output quality indicator, directly influencing paste rheology and pipeline transport performance. Target concentrations are typically high (>65% wt solids). Online densitometers (e.g., nuclear density gauges, ultrasonic concentration meters) are susceptible to variations in slurry physical properties (particle size distribution, mineral composition), air bubbles, sensor coating, wear, and the thickener's complex internal flow dynamics. This leads to fluctuating measurements, reduced accuracy, or even failure. Maintenance costs are high, and reliability under extreme operating conditions is often insufficient.
- **Inflow/feed concentration:** Fluctuations in upstream mineral processing cause significant variations in inflow concentration and flow rate, acting as major disturbances for the thickener. Although inflow concentration is generally lower than underflow concentration and slightly easier to measure, it still faces challenges such as slurry variability and instrument calibration drift. Accurate knowledge of the inflow

state is vital for feedforward control and understanding thickener loading.
- **Mud bed height/sludge level:** This is a core process variable representing the internal solid-phase inventory, settling/compression state, and operational stability of the thickener. An optimal mud bed height is key to ensuring clarification effectiveness (overflow water quality) and underflow concentration. Direct measurement is extremely difficult. Mechanical mud level indicators (e.g., inferred from rake torque, floats) suffer from response lag and are prone to mechanical failure and structural interference. Non-contact instruments (e.g., radar, ultrasonic) experience severe signal attenuation due to steam, foam, splashing, and dust near the thickener top, compromising measurement accuracy and reliability.
- **Underflow discharge rate:** Although flow meters are a relatively mature technology, conventional types (e.g., electromagnetic, Coriolis mass flow meters) are prone to clogging, abrasion, and measurement deviations under the demanding conditions of discharging high-density, high-viscosity, solids-laden paste that may contain entrained air. This necessitates frequent maintenance.

These key quality variables (KQVs) exhibit strong **coupling relationships**:
- Changes in the mud bed height directly impact the stability of the underflow concentration and the pressure in the compression zone.
- Fluctuations in inflow concentration and flow rate rapidly disturb the mud bed height.
- The underflow discharge rate is the primary manipulated variable for regulating both mud bed height and underflow concentration, yet its adjustment is constrained by the current mud bed state and underflow concentration target.
- The effectiveness of flocculant addition is also highly dependent on inflow properties and the mud bed state.

Given the **high cost, low reliability, maintenance difficulties, or technical limitations** associated with direct online measurement of these critical KQVs, combined with their **decisive role** in paste quality and process stability, and the **complex dynamic coupling** between them, the development and application of **soft sensing techniques** for deep cone thickeners becomes a critical and urgent task.

The core task of soft sensing for deep cone thickeners is to construct a mathematical model (soft sensor) capable of **estimating in real-**

time, accurately, and reliably the aforementioned difficult-to-measure KQVs (primarily **underflow concentration, inflow concentration, and mud bed height**). This model is built using:

- Easily and reliably obtainable auxiliary process variables (e.g., inflow rate, overflow rate, underflow rate, rake drive torque/current, flocculant dosage, pressure at various points, temperature).
- Mechanistic knowledge reflecting the physicochemical characteristics of the process (e.g., settling kinetics, mass conservation).
- Advanced data-driven modeling (e.g., machine learning, deep learning) or hybrid modeling approaches.

These soft-sensed values serve as crucial process state information, providing decision support for operators and acting as direct inputs to advanced process control (APC) systems, such as model predictive control (MPC). This enables more precise, stable, and optimized control of the deep cone thickener, ultimately ensuring the safe, efficient, and economical operation of the paste-filling process.

4.3 Conclusions

The paste-filling system in underground mining presents numerous complex prediction and soft sensing challenges due to its multivariable coupled dynamics, harsh operating conditions, and stringent quality requirements. Data-driven machine learning approaches offer powerful and innovative solutions to address these challenges, enabling accurate estimation of critical process variables that are otherwise difficult to measure directly [14].

This chapter has established the fundamental context of paste-filling operations, highlighting their significance in modern mining. We have identified and analyzed key prediction and soft sensing tasks essential for optimizing thickener performance and ensuring paste quality. Particular emphasis was placed on the estimation of critical quality variables such as underflow concentration, feed concentration, and mud bed height, all crucial for stable process control.

Building upon this foundation, the subsequent chapter introduces fundamental machine learning architectures forming the basis for effective time-series modeling. These methodologies are presented with specific focus on their applicability to the dynamic, non-linear systems characteristic of industrial paste-filling processes.

References

[1] Yongxiang Lei, Hamid Reza Karimi, Underflow concentration soft sensing for cone thickener system based on a direct data-driven quantile regression forecasting, Mechatronics Technology 2 (1) (2025) 19–20.
[2] Yongxiang Lei, Hamid Reza Karimi, Underflow concentration prediction based on improved dual bidirectional LSTM for hierarchical cone thickener system, The International Journal of Advanced Manufacturing Technology (2023) 1–12.
[3] Zhaolin Yuan, Xiaorui Li, Di Wu, Xiaojuan Ban, Nai-Qi Wu, Hong-Ning Dai, Hao Wang, Continuous-time prediction of industrial paste thickener system with differential ODE-Net, IEEE/CAA Journal of Automatica Sinica 9 (4) (2022) 686–698, https://doi.org/10.1109/JAS.2022.105464.
[4] Hua zhe Jiao, Yachuang Wu, Hui Wang, Xinming Chen, Zhen Li, Yunfei Wang, Boyang Zhang, Juanhong Liu, Micro-scale mechanism of sealed water seepage and thickening from tailings bed in rake shearing thickener, Minerals Engineering (2021), https://doi.org/10.1016/j.mineng.2021.107043.
[5] Ji-Ning Xu, Zhan-Bin Zhao, Feng-Qiang Wang, Data driven control of underflow slurry concentration in deep cone thickener, in: 2017 6th Data Driven Control and Learning Systems (DDCLS), 2017, pp. 690–693.
[6] Dong Xiao, Hongfei Xie, Longqiang Jiang, Ba Tuan Le, Jichun Wang, Chongmin Liu, Hongzong Li, Research on a method for predicting the underflow concentration of a thickener based on the hybrid model, Engineering Applications of Computational Fluid Mechanics 14 (2020) 13–26, https://doi.org/10.1080/19942060.2019.1658228.
[7] A. Wu, Z. Ruan, Cui ping Li, Shaoyong Wang, Yong Wang, Jiandong Wang, Numerical study of flocculation settling and thickening of whole-tailings in deep cone thickener using CFD approach, Journal of Central South University 26 (2019) 711–718, https://doi.org/10.1007/S11771-019-4041-7.
[8] Yongxiang Lei, Intelligent optimal framework for the industrial mining plant-wide prediction control, IEEE Transactions on Instrumentation and Measurement (2023).
[9] Zhaolin Yuan, Jinlong Hu, Di Wu, X. Ban, A dual-attention recurrent neural network method for deep cone thickener underflow concentration prediction, Sensors 20 (2020), https://doi.org/10.3390/s20051260.
[10] Yongxiang Lei, Hamid Reza Karimi, DualLSTM: a novel key-quality prediction for a hierarchical cone thickener, Control Engineering Practice 137 (2023) 105566.
[11] Xiaorui Li, Xiaojuan Ban, Yizhou Shao, Haowei Chen, Chao Yao, Zhaolin Yuan, Guangsheng Liu, Lijie Guo, Process industry system modeling based on time series imputation: a case study of paste thickener, in: 2024 International Conference on Automation in Manufacturing, Transportation and Logistics (ICaMaL), 2024, pp. 1–5.
[12] Yongxiang Lei, Hamid Reza Karimi, Lihui Cen, Xiaofang Chen, Yongfang Xie, Processes soft modeling based on stacked autoencoders and wavelet extreme learning machine for aluminum plant-wide application, Control Engineering Practice 108 (2021) 104706.
[13] Chunkang Liu, Hongjiang Wang, Aixiang Wu, Bolin Xiao, Jun Nie, Min Liu, Evolution laws of underflow concentration and rake torque during initial commissioning of DCT: insight into the impact of high-level circulation and floc residence time, in: Construction and Building Materials, 2025.
[14] Sarah Almaghrabi, Mashud Rana, Margaret Hamilton, Mohammad Saiedur Rahaman, Multivariate solar power time series forecasting using multilevel data fusion and deep neural networks, Information Fusion 104 (2024) 102180.

PART 2

Methodology

CHAPTER 5

Machine learning basics for prediction and soft sensing

5.1 Deterministic models

5.1.1 ARIMA (autoregressive integrated moving average)

ARIMA is a time series forecasting model with wide applications such as stock prediction [1] and electricity price [2]. A autoregression diagram is given in Fig. 5.1. In ARIMA the historical information is formulated in linear combination [3]. The objective developed is to predict the future $\mathbf{x_t}$ based on the historical time series $\mathbf{x_{t-1}}$. The AR approach is built on the assumption that $\mathbf{x_t}$ at time instant t is a linear approximation of its past observation $F(t-1), F(t-2), \ldots, F(t-m), F(t-m-1)$:

$$\begin{aligned} &F(t) = a_1 F(t-1) + a_2 F(t-2) \\ &+ \cdots + a_m F(t-m) + a_{m+1} F(t-m-1) + v(t), \end{aligned} \tag{5.1}$$

where $(a_1, a_2, \ldots, a_{m+1})$ are the parameters, $m+1$ is the order of the autoregressive model, and $v(t)$ is the zero-mean white noise distribution. Suppose that the coefficients a_i, $i = 1, 2, 3, \ldots, m, m+1$, at time t are already calculated. So the best prediction of $\hat{F}(t+T|t)$ at time t is

$$\begin{aligned} &\hat{F}(t+T|t) = a_1(t)F(t+T-1|t) \\ &+ a_2(t)F(t+T-2|t) + \cdots + a_m(t)F(t+T-m|t) \\ &+ a_{m+1}(t)F(t+T-m-1|t), \end{aligned} \tag{5.2}$$

where $\hat{F}(t+T-i|t) = F(t+T-i)$ if $t+T-i \leq t$ or $t-i \leq 0$.

Eq. (5.1) can be rewritten in a compact vector form

$$\mathbf{x_t} = \mathbf{x_{t-1}}^T \mathbf{A} + \mathbf{v(t)}, \tag{5.3}$$

where $\mathbf{A} = [a_1, a_2, \ldots, a_{m+1}]^T$, and

$$\mathbf{x_{t-1}} = [F(t-1), F(t-2), \ldots, F(t-m), F_{exc}(t-m-1)]^T.$$

The solution target of (5.1) is to obtain the coefficient vector $\mathbf{A}_i$, $i = 1, 2, \ldots, m+1$, at time t. It can be obtained by minimizing the following

https://doi.org/10.1016/B978-0-44-336759-5.00013-5

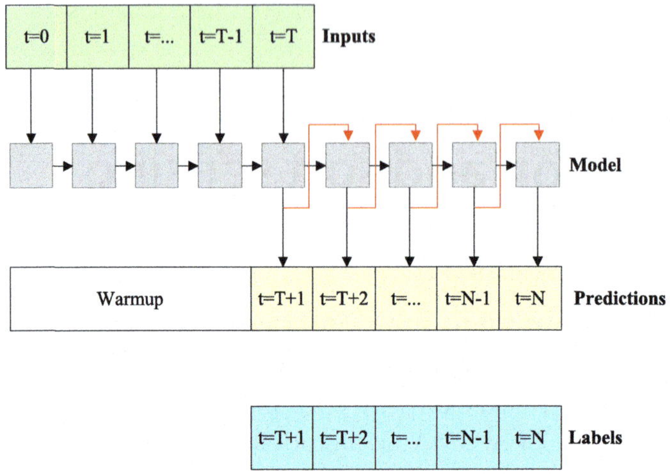

Figure 5.1 Structure of autoregression.

multi-step long-range accumulation ahead error:

$$\min_{a_1, a_2, \ldots, a_{m+1}} \sum\nolimits_{i=m+N_p+1}^{t} \sum\nolimits_{i=1}^{N_p} \left(F(t) - \hat{F}(t|t-i)\right)^2, \tag{5.4}$$

where N_p is the forecasting horizon over the AR model. After the optimization training process, the AR-based model can predict the future N_p steps based on the historical $m+1$ past observation steps.

5.1.2 ANN (artificial neural network)

The realm of artificial intelligence has undergone a remarkable transformation over the past few decades, catalyzed by the inception and evolution of neural networks [4,5]. As computational architectures inspired by the intricate network of interconnected neurons in the human brain, neural networks have emerged as a transformative paradigm for tackling complex problems in various domains [6,7]. This thesis delves into the multifaceted world of neural networks, exploring their architecture, training methodologies, applications, and the profound impact they have had on the fields of machine learning, data analysis, and decision making.

At the heart of neural networks lies their architecture, characterized by layers of interconnected artificial neurons, known as nodes or units (see Fig. 5.2). These neurons mimic biological neurons in their ability to process and transmit information. Neural network architectures encompass a spectrum of designs, from basic feedforward networks to more intricate convolutional, recurrent, and attention-based networks. The interplay of input, hidden, and output layers forms the basis for information processing, enabling these networks to extract intricate patterns, relationships, and features from complex datasets. It also mines the information in our process industry tasks.

Central to the effectiveness of neural networks is their capacity to learn from data. This learning process is facilitated through training, where the network iteratively adjusts its internal parameters to minimize the error between predicted and actual outcomes [8]. Supervised learning involves pairing input data with corresponding target outputs, whereas unsupervised learning uncovers underlying patterns in unlabeled data. Reinforcement learning, on the other hand, guides the network behavior through rewards and punishments. Optimization of network parameters is achieved through gradient-based techniques such as backpropagation, which allows the network to converge to optimal solutions.

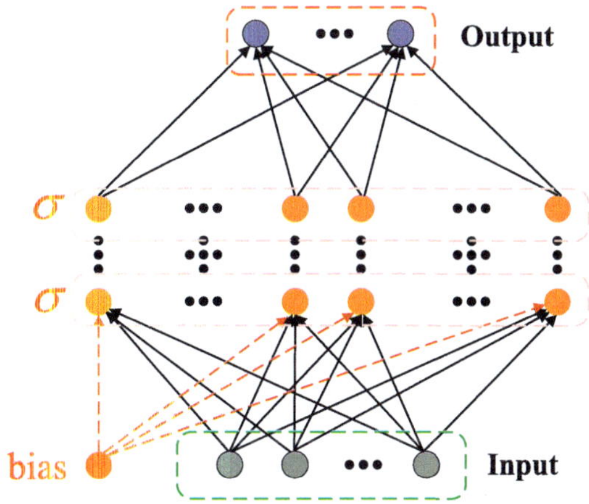

Figure 5.2 An architecture of neural networks.

The versatility of neural networks extends across a myriad of applications, spanning image and speech recognition, natural language processing, autonomous driving, healthcare diagnostics, finance, and beyond. The ability to handle massive datasets, adapt to complex input patterns, and generalize from limited examples has revolutionized fields previously deemed impervious to computational solutions. Neural networks have not only achieved unprecedented accuracy in classification, prediction, and decision-making tasks but also have spawned innovations in generative modeling, transfer learning, and interpretability.

Although the potential of neural networks is undeniable, challenges such as over-fitting, interpretability, and bias mitigation remain pivotal. Moreover, the growth of deep learning, a subset of neural networks comprising multiple hidden layers, has brought computational complexity and resource requirements to the forefront. As research continues, innovations such as adversarial training, explainable AI, and self-attention mechanisms hold promise for addressing these challenges and unlocking new frontiers in neural network research.

In this book, we embark on a comprehensive exploration of neural networks, unraveling their architectural intricacies, elucidating learning paradigms, and dissecting their profound applications in industrial soft sensing and prediction tasks. By delving into the nuances of their operation, assessing their limitations, and envisioning future trajectories, this study seeks to contribute to the broader discourse on the transformative power of neural networks in reshaping the landscape of artificial intelligence and decision support systems.

5.1.3 RNN (recurrent neural network)

Recurrent neural networks (RNNs) are a class of artificial neural networks designed for processing sequential data by maintaining a hidden state that captures information from previous time steps [9–11] and enabling the modeling of temporal dependencies [12], see Fig. 5.3.

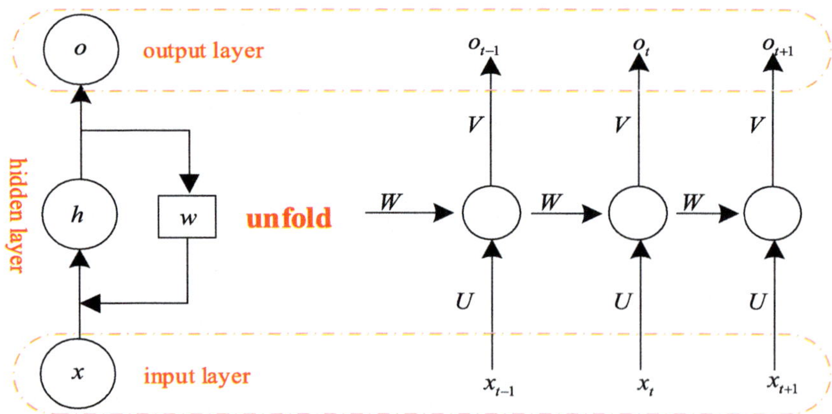

Figure 5.3 The structure of RNN.

Given an input sequence $\{x^{\langle 1\rangle}, x^{\langle 2\rangle}, \ldots, x^{\langle T\rangle}\}$, the hidden state $h^{\langle t\rangle}$ and output $y^{\langle t\rangle}$ at time step t are computed as

$$h^{\langle t\rangle} = \phi\left(W_{hh} h^{\langle t-1\rangle} + W_{xh} x^{\langle t\rangle} + b_h\right), \qquad y^{\langle t\rangle} = \psi\left(W_{hy} h^{\langle t\rangle} + b_y\right),$$

where W_{hh}, W_{xh}, and W_{hy} are weight matrices, b_h and b_y are biases, and ϕ and ψ are activation functions, such as `tanh`, ReLU, or `softmax`.

RNNs are trained with backpropagation through time (BPTT), which unfolds the network over time and applies standard backpropagation to the extended computational graph [13,14]. During BPTT, gradients may vanish or explode, often mitigated by techniques such as gradient clipping [15].

To address vanishing/exploding gradients and improve memory capacity, several gated variants have been proposed:

- **Long short-term memory (LSTM)** [16] introduces input, forget, and output gates.
- **Gated recurrent unit (GRU)** [17] combines gates to reduce parameters while retaining performance.
- **Bidirectional RNN (Bi-RNN)** [18] processes sequences in both forward and backward directions.

RNNs and their variants are widely used in:

- Natural language processing: language modeling, machine translation, text generation [19].
- Time series forecasting: stock prices, weather, sensor data [20].
- Speech and audio: speech recognition, emotion detection, music generation [21].
- Video analysis: sequence-to-sequence video captioning [22].

5.1.4 LSTM (long short-term memory)

The LSTM [16] copes with a long dependency by inducing different gates so that useful information can be reserved for the next layer [23,24]. As shown in Fig. 5.4, the main equation for information learning can be given as

$$\hat{y} = \sigma\left(V \cdot h + b\right), \tag{5.5}$$

where $\hat{y}$ is the estimated output value, σ is the activation function, and h and b are the hidden layer and bias, respectively.

Several experiments and applications have shown the superior capability of LSTM. Furthermore, many other extensive architectures such as deep

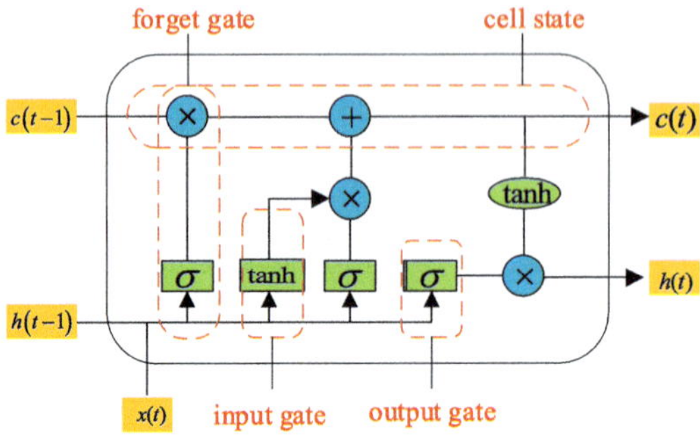

Figure 5.4 An architecture of LSTM.

LSTM, have been used in many domains. A widely accepted shortcoming of RNN is the problem of long-term dependency and vanishing gradients. A special kind of RNN called LSTM is capable of learning long-term dependencies [25,26]. LSTM was introduced by Hochreiter and Schmidhuber and is widely used in a wide variety of problems. In the basic LSTM, three gates are added to protect and control the cell state, and some repeating modules are added to an LSTM that contains four interacting layers. The LSTM network can implement temporal memory through the switch of the gates to prevent gradient varnishing. Denote the input vector by $x(t)$, the previous hidden state by $h(t-1)$, and the external input, its previous cell state, by $c(t-1)$. Then the forget gate is triggered as

$$f(t) = \sigma\left(W_f \cdot \left[h_{t-1}, x_t\right] + b_f\right), \tag{5.6}$$

where $\cdot$ means the dot product, and the input gate and new candidate vectors are computed as

$$i_t = \sigma\left(W_i \cdot \left[h_{t-1}, x_t\right] + b_i\right), \tag{5.7}$$

$$\tilde{C}_t = \tanh\left(W_{(C)} \cdot \left[h_{t-1}, x_t\right] + b_C\right). \tag{5.8}$$

The new cell state in LSTM is updated by the equation

$$C_t = f_t * C_{t-1} + i_t * \tilde{C}_t. \tag{5.9}$$

The output gate vector can be calculated as follows:

$$o_t = \sigma\left(W_o\left[h_{t-1}, x_t\right] + b_o\right), \tag{5.10}$$

$$h_t = o_t * \tanh\left(C_t\right) / \tag{5.11}$$

In Eqs. (5.6)–(5.11), σ is the nonlinear activation function, usually, the *sigmoid* function, tanh represents the nonlinear tangent activation function, $*$ represents the point-wise multiplication operation, W_c, W_o, W_i are the related weights, and b_i, b_c, b_o are the biases, respectively.

Compared to the traditional RNN, LSTM maintains a cell state to accumulate the long-time feature from the time sequel [27]. So the derivatives of the features from the front instant would be preserved. Furthermore, to guarantee the performance of temperature prediction, the information derivative over the time axis is very important, which promotes the proposed method in the sequel [28].

5.1.5 GRU (gated recurrent unit)

The gated recurrent unit (GRU) was introduced by Cho et al. [17] as a streamlined alternative to LSTM, combining the memory cell and gates into a single hidden state while retaining the ability to capture long-term dependencies. At each time step t, the GRU uses two gates, the update gate $z^{\langle t \rangle}$ and reset gate $r^{\langle t \rangle}$, to control information flow [29]. The gate activations are computed as

$$z^{\langle t \rangle} = \sigma\left(W_z x^{\langle t \rangle} + U_z h^{\langle t-1 \rangle} + b_z\right), \quad r^{\langle t \rangle} = \sigma\left(W_r x^{\langle t \rangle} + U_r h^{\langle t-1 \rangle} + b_r\right),$$

where σ is the sigmoid function [30]. A candidate activation $\tilde{h}^{\langle t \rangle}$ is then formed using the reset gate as

$$\tilde{h}^{\langle t \rangle} = \tanh\left(W_h x^{\langle t \rangle} + U_h \left(r^{\langle t \rangle} \odot h^{\langle t-1 \rangle}\right) + b_h\right),$$

and the new hidden state is a convex combination of the previous state and candidate:

$$h^{\langle t \rangle} = z^{\langle t \rangle} \odot h^{\langle t-1 \rangle} + \left(1 - z^{\langle t \rangle}\right) \odot \tilde{h}^{\langle t \rangle}.$$

This design both mitigates the vanishing gradient problem and reduces computational complexity compared to LSTM.

GRUs are trained via backpropagation through time (BPTT) in the same manner as standard RNNs, with gating helping stabilize gradients and often requiring fewer parameters and faster convergence than LSTM-based models [31,32]. Empirical evaluations have demonstrated that GRUs achieve comparable performance to LSTMs on tasks like polyphonic music modeling and speech signal modeling [33,34], and they have been successfully applied in machine translation [35,34], speech recognition [36], and time series forecasting [37].

5.1.6 CNN (convolutional neural network)

The structure of convolutional neural networks is shown in Fig. 5.5. They stand as a pivotal achievement in the evolution of deep learning, offering unparalleled capabilities in visual data analysis. The architectural nuances of CNNs are specifically tailored to address the challenges posed by image and spatial data, propelling their widespread adoption across diverse domains [38]. At the heart of CNNs lies a distinctive architecture that draws inspiration from the organization of the animal visual cortex. Characterized by convolutional layers, these networks excel at local feature extraction. The convolutional layers are complemented by pooling layers, which efficiently downsample spatial dimensions and fully connected layers for high-level reasoning. A defining feature of CNNs is their ability to

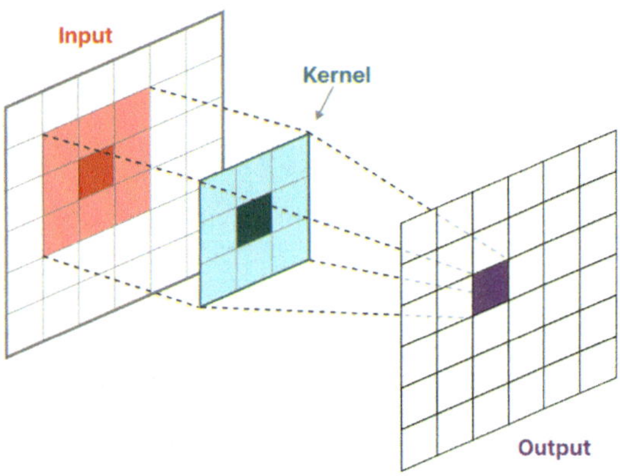

Figure 5.5 An architecture of CNN.

autonomously learn hierarchical representations of features. As the data travel through successive layers, the network progressively abstracts intricate features, proving instrumental in tasks such as image classification, object detection, and segmentation. If we deal with text rather than images, then we can consider the one-dimensional perspective instead (Fig. 5.6): The impact of CNNs resonates across various applications, demonstrating superiority in image recognition, medical imaging, autonomous systems, and natural language processing. The capacity to generalize complex

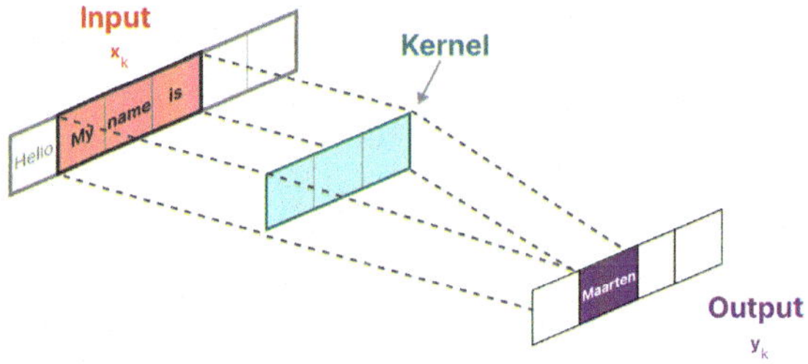

Figure 5.6 The structure of one-dimensional convolution.

patterns from extensive datasets positions CNNs at the forefront of state-of-the-art performance in benchmark tasks [39].

As we delve deeper into the intricacies and applications of CNNs in subsequent sections, this segment serves to illuminate the foundational aspects of their architecture and underscores their transformative impact on the landscape of deep learning.

5.1.7 AE (auto-encoder)

Auto-encoder (AE) is an unsupervised single-hidden-layer neural network that maps the output layer equal to the input layer as accurately as possible [40]. AE constructs a three-layer neural network to fine-tune the optimal parameters. Unlike the radial basis network, the output of AE can be the same as the input, see Fig. 5.7. Assume that the input dimension of the

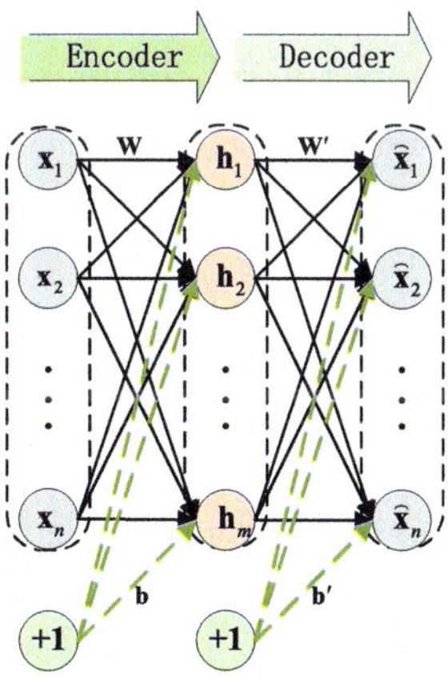

Figure 5.7 An architecture of AE.

autoencoder is $\mathbf{x} = [x_1, x_2, \ldots, x_{n_i}]^T$, where n_i is the dimension of the input variables. First, the auto-encoder projects from the input layer to the hidden layer by the mapping function as follows:

$$\mathbf{H} = f(\mathbf{W}\mathbf{x} + \mathbf{b}), \tag{5.12}$$

where $\mathbf{H} = [h_1, h_2, \ldots, h_{n_h}]^T$ is the hidden representation, n_h is the dimension of hidden neurons, $\mathbf{W}$ is a weight index matrix of dimension of $n_i \times n_h$, and $\mathbf{b}$ is the bias vector. The cost function of an auto-encoder can be expressed as

$$L_{AE} = \frac{1}{N}\sum_{i=1}^{N} |x_i - \hat{x}_i|^2 = \frac{1}{N}\sum_{i=1}^{N} |x_i - g(a \cdot x_i + b)|^2, \tag{5.13}$$

where N is the input, and g is the related activation function. The loss function is solved by a gradient descent algorithm. Some variety of AEs has been built, such as a denoising and variational auto-encoders. To obtain deep feature representation and dimension reduction, the basic auto-encoder can be overlapped layer-by-layer.

5.1.8 Transformer models

Transformer is a novel encoder–decoder network architecture based on a self-attention mechanism. Although Transformer was originally applied to sequence-to-sequence learning for language understanding, it has now been extensively applied in various fields, such as vision, speech, time series, and reinforcement learning. Due to the superiority of its self-attention mechanism, Transformer has demonstrated powerful modeling capabilities for capturing correlations and interactions in time-series data. In recent years, many Transformer variants have been proposed to tackle various time-series tasks and have been used for time-series forecasting [41], anomaly detection [42], classification [43], etc., achieving strong performance [44], see Fig. 5.8.

In the case of time series forecasting, the general structure of Transformer is shown in Fig. 5.6. The encoder receives the input of the time series (i.e., X_{en}), and the input of the decoder comes from the partial sequence of X_{en} and the placeholder padding of the forecast target elements. Furthermore, the inputs of the encoder and decoder embed the position encoding information of the time series and some covariates (e.g., timestamp information). The output of the decoder is the target value of the forecast element.

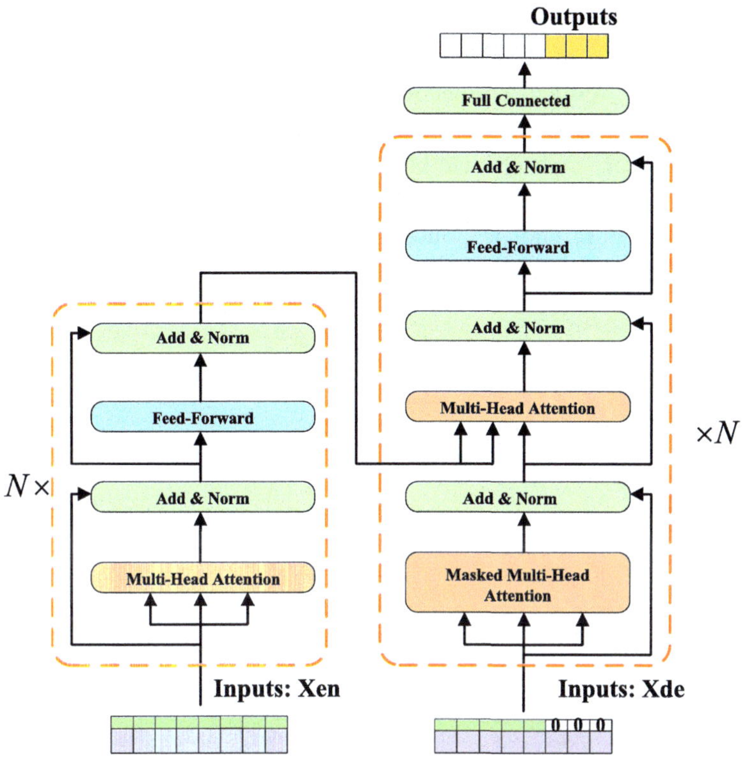

Figure 5.8 Typical structure of Transformer.

5.1.9 TCN (temporal convolutional network)

As shown in Fig. 5.10, a TCN is a general convolutional architecture suitable for sequence modeling. Unlike traditional CNNs, TCNs use causal convolution to ensure that the output at time t depends only on the input values at time t and earlier [45]. Furthermore, TCNs address the limitation of the receptive field by introducing dilated convolution. Generally, for a one-dimensional sequence input $x \in \mathbb{R}^n$, the dilated convolution output at location t can be expressed as

$$s(t) = \sum_{i=0}^{k-1} f_i \cdot x_{t-d \cdot i}, \tag{3}$$

where d is the dilation factor, k is the kernel size, and f_i represents the ith convolution kernel. Fig. 5.9 shows an example of the dilated causal convolution with dilation factors $d = \{1, 2, 4, 8\}$ and kernel size $k = 2$, which can cover a receptive field of size 16 by stacking four layers.

In addition, residual connections are used in TCNs to accelerate the feedback and convergence of deep networks. Fig. 5.10 shows the residual block of a TCN. Each residual block contains two layers of dilated causal convolution, where the weights of the convolution kernels are weight-normalized. ReLU and dropout are applied after each convolution, and downsampling ensures that the input and output additions are compatible.

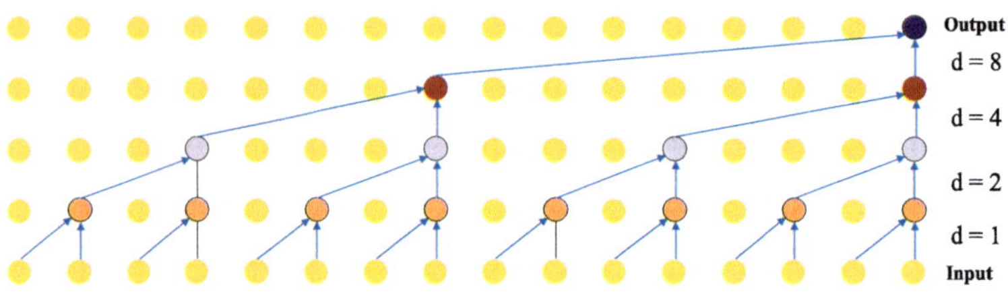

Figure 5.9 The dilated causal convolution.

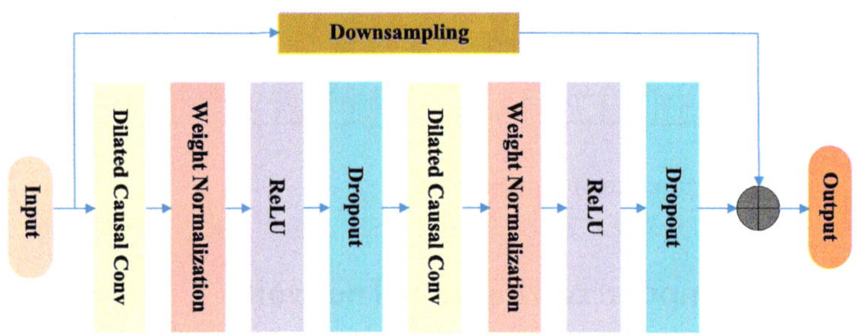

Figure 5.10 A TCN residual block.

5.1.10 ELM (extreme learning machine)

ELM is proposed in [4], which has been considered an efficient and effective method to train the single hidden-layer feedback neural network. The core idea in ELM is that a Moore–Penrose solution is used as an alternative to the gradient descent process. It is proved that ELM has fast computation and high generalization ability. Given a set of training samples $\{\mathbf{x}, \mathrm{t}\}_{i=1}^{N}$, ELM constructs a network with two stages (Fig. 5.11). In the first stage, randomly generated input weights are sampled from a uniform distribution in $(-1, 1)$. The bias b is initialized by a constant, so the output

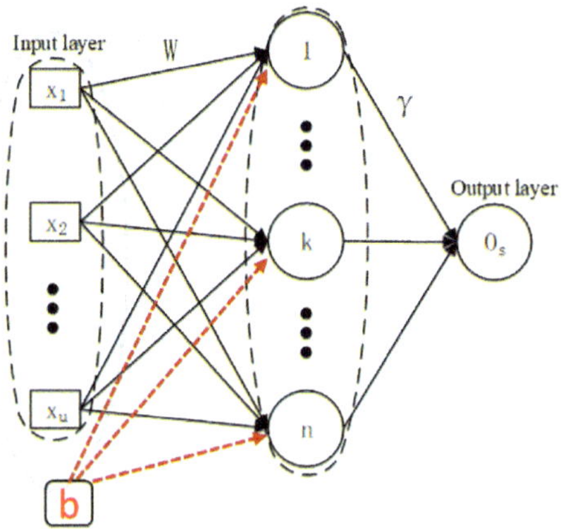

Figure 5.11 An architecture of ELM.

of the hidden layer is directly represented as

$$H = \begin{pmatrix} G(w_1, b_1, x_1) & \dots & G(w_1, b_L, x_N) \\ \vdots & \ddots & \vdots \\ G(w_1, b_1, x_N) & \cdots & G(w_1, b_L, x_N) \end{pmatrix}, \tag{5.14}$$

where n is the number of neurons in the hidden layer of the neural network, G is the activation kernel function, which is usually a nonlinear piecewise function. The single hidden-layer feedforward neural network can be described as

$$\sum_{\mathrm{i}=1}^{N} \gamma g\left(w\mathbf{x}_i + b\right) = \mathbf{y}_i \tag{5.15}$$

with activation function g. The learning approximation function with N samples is

$$\min_{\gamma} \left\| \sum_{\mathrm{i}=1}^{N} \gamma g\left(w\mathbf{x}_i + b\right) - \mathbf{y}_i \right\|. \tag{5.16}$$

So a commonly least square objective function can be built by the equation

$$\mathbf{H}\gamma = \mathbf{Y} \tag{5.17}$$

Its output weight solution can be given in the following Moore-inverse formulation:

$$\gamma = \mathbf{H}^{\dagger} \cdot \mathbf{Y}, \tag{5.18}$$

which is a closed solution of the hidden weight β rather than an iterative training, so the time complexity will be greatly improved.

5.1.11 GNN (graph neural network)

Graph neural networks (GNNs) are a class of deep learning models designed to process data structured as graphs, where entities (nodes) and their relationships (edges) are explicitly encoded. Unlike traditional neural networks, which assume Euclidean data structures (e.g., grids or sequences), GNNs operate on non-Euclidean domains, enabling them to capture complex relational patterns and dependencies inherent in real-world systems. This capability makes GNNs particularly powerful for tasks involving social networks, molecular structures, recommendation systems, and—increasingly—**multivariate time series prediction** with spatial or relational dependencies, see Fig. 5.12.

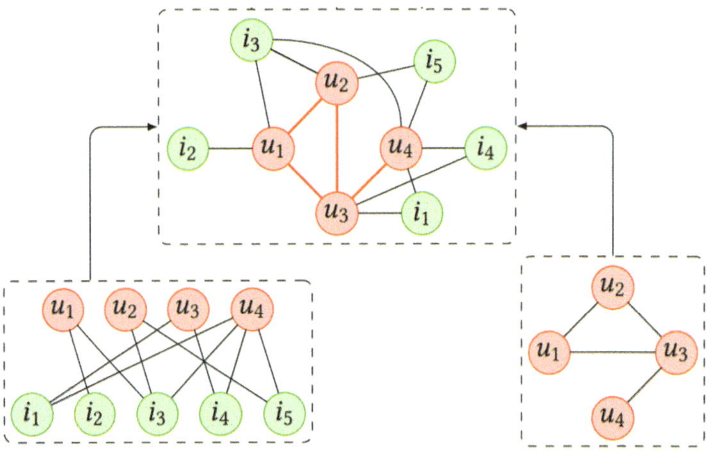

Figure 5.12 Structure of GNNs.

The foundational principle of GNNs lies in *message passing*, a mechanism that iteratively updates node representations by aggregating information from neighboring nodes. This process mimics human reasoning about relational contexts, where the state of an entity is influenced by its interactions with others. Formally, a GNN layer can be decomposed into two stages:

1. **Aggregation**: Each node gathers features from its local neighborhood. For node v, the aggregated message $m_v^{(l)}$ at layer l is computed as

$$m_v^{(l)} = \mathrm{AGGREGATE}^{(l)}\left(\left\{h_u^{(l-1)} \mid u \in \mathcal{N}(v)\right\}\right), \tag{5.19}$$

   where $h_u^{(l-1)}$ denotes the hidden state of neighbor u at layer $l-1$, and $\mathcal{N}(v)$ is the set of nodes adjacent to v. Common aggregation functions include summation, mean, max-pooling, or attention-weighted combinations.

2. **Update**: The node integrates the aggregated message with its previous state to generate a refined representation:

$$h_v^{(l)} = \mathrm{UPDATE}^{(l)}\left(h_v^{(l-1)}, m_v^{(l)}\right). \tag{5.20}$$

   The update function often employs learnable transformations, such as linear layers followed by nonlinear activations (e.g., ReLU) or gated mechanisms (e.g., GRU).

By stacking multiple GNN layers, nodes progressively incorporate information from broader neighborhoods, enabling the model to capture both local and global graph structures.

Several GNN architectures have been developed to enhance aggregation efficiency and expressive power:

- **Graph convolutional network (GCN)** [46] simplifies spectral graph convolutions using a first-order approximation of Chebyshev polynomials. The layer-wise propagation rule is

$$H^{(l+1)} = \sigma\left(\hat{D}^{-\frac{1}{2}}\hat{A}\hat{D}^{-\frac{1}{2}}H^{(l)}W^{(l)}\right), \tag{5.21}$$

  where $\hat{A} = A + I$ (graph with self-loops), $\hat{D}$ is the degree matrix of $\hat{A}$, and $W^{(l)}$ is a learnable weight matrix.

- **Graph attention network (GAT)** [47] introduces attention mechanisms to dynamically assign weights to neighbors during aggregation:

$$h_v^{(l)} = \sigma\left(\sum_{u \in \mathcal{N}(v)} \alpha_{vu}^{(l)} W^{(l)} h_u^{(l-1)}\right), \tag{5.22}$$

  where α_{vu} is the attention coefficient between nodes v and u, computed via a softmax over learned similarity scores.

- **GraphSAGE** [48] extends GNNs to inductive settings by sampling fixed-size neighborhoods and employing parameterized aggregators (e.g., LSTM, mean):

$$h_v^{(l)} = \sigma\left(W^{(l)} \cdot \text{CONCAT}\left(h_v^{(l-1)}, \text{AGGREGATE}^{(l)}\left(\{h_u^{(l-1)}\}\right)\right)\right). \tag{5.23}$$

- **Temporal graph networks (TGNs)**: For dynamic graphs, TGNs [49] incorporate temporal embeddings by updating node states based on both structural and temporal interactions.

In time series forecasting, GNNs excel at modeling spatio-temporal dependencies, leveraging both historical trends (temporal) and cross-variable relationships (spatial). For example, in traffic flow prediction, road sensors (nodes) exhibit dependencies influenced by network topology (edges), whereas their readings evolve over time.

- **Spatio-temporal coupling**, joint modeling via stacked architectures:

$$\mathcal{H}_{t+1:t+\tau} = f_{\text{GNN}}(f_{\text{Temporal}}(\mathbf{X}_{t-k:t})). \tag{5.24}$$

- **Dynamic graph construction**, adaptive adjacency matrix learning:

$$A_{\text{adaptive}} = \text{SoftMax}(\text{ReLU}(E_1 E_2^\top)), \tag{5.25}$$

  where E_1 and E_2 are learnable node embeddings.
- **Strengths**:
  - Captures cross-variable dependencies
  - Handles irregular sampling intervals
  - Enables zero-shot transfer learning
- **Limitations**:
  - $\mathcal{O}(N^2)$ memory complexity
  - Sensitive to graph construction
  - Theoretical interpretability gaps

5.1.12 Mamba (state-space sequence model)

Mamba is a novel sequence modeling architecture based on the selective state space model (SSSM), proposed by Gu and Dao [50] in 2023. By combining a linear-time computational framework with context-aware modeling capabilities, it overcomes the quadratic complexity bottleneck of Transformers in long sequence processing, achieving outstanding performance in fields such as natural language processing, genomics, and time series forecasting [50,51], see Figs. 5.13 and 5.14.

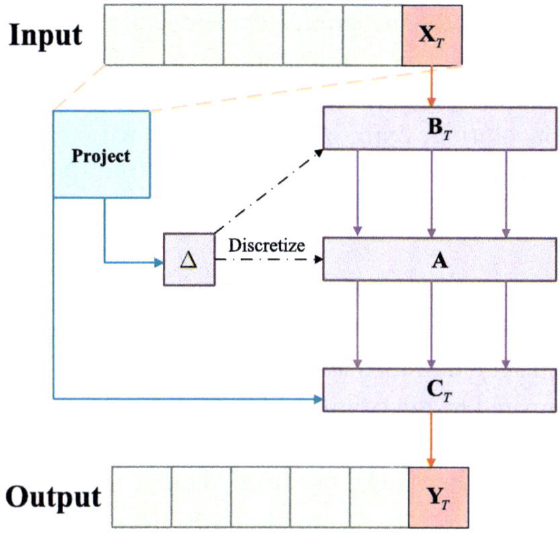

Figure 5.13 Selective state space.

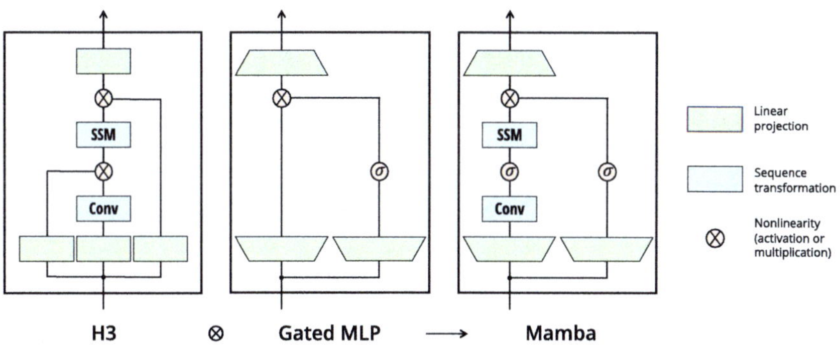

Figure 5.14 Structure of Mamba sourced from [50].

The core innovation of Mamba lies in introducing input-dependent dynamics into traditional state space models (SSMs), enabling "selective" sequence modeling:

1. **Selective state space**: SSM coefficients (Δ, B, C) are parameterized by the input to dynamically adjust the information propagation strategy:

$$h_t = A(\Delta(x_t)) \odot h_{t-1} + B(\Delta(x_t)) \odot x_t, \quad (5.26)$$

where $\Delta(x_t)$ denotes the input-dependent discretization step, and $\odot$ represents element-wise multiplication.

2. **Hardware-aware algorithms**: Mamba employs scan operations instead of convolution, optimizing computational efficiency by leveraging the hierarchical memory structure of GPUs. On an A100 GPU, it achieves up to 3× faster training speed compared to traditional SSMs [50,52].

- **Structured state space duality (SSD)**: Mamba-2 unifies SSM and linear attention through block matrix decomposition, enabling semi-separable matrix transformations and achieving 2–8× faster training speeds compared to the original Mamba.
- **Hybrid architecture design**: Mamba constructs a homogeneous architecture by alternately stacking Mamba layers with gated MLP blocks, removing traditional attention mechanisms:

$$\text{MambaBlock}(x) = \text{MLP}(\text{SSM}(x) \otimes \text{GeLU}(x)), \tag{5.27}$$

where $\otimes$ denotes a gating mechanism [50,51].

- **Spatiotemporal modeling framework**: Models such as TimeMachine employ a quadruple Mamba architecture to handle multivariate sequences:

$$\hat{Y}_{t+1:t+\tau} = \sum_{i=1}^{4} \text{Mamba}_i(\text{DownSample}(X_{t-k:t})), \tag{5.28}$$

capturing both global and local dependencies through multi-scale downsampling.

- **Bidirectional modeling**: The Bi-Mamba+ model introduces a forget-gate mechanism, combining past and future contexts:

$$h_t^{\text{bi}} = [\text{Mamba}_{\rightarrow}(x_{1:t}); \text{Mamba}_{\leftarrow}(x_{t:T})], \tag{5.29}$$

achieving a 23% reduction in MAE on the ETTh2 dataset, see Table 5.1.

- **Advantages**:
  - Linear computational complexity ($O(L)$ vs. Transformer's $O(L^2)$) [50,51]
  - Supports million-length sequence modeling
  - Inference throughput up to 5× that of Transformers

Table 5.1 Comparison between Mamba and Mainstream architectures.

Model	Time Complexity	Maximum Sequence Length	GPU Memory Usage
Transformer	$O(L^2)$	32k	High
GNN	$O(N+E)$	Dynamic Graph	Medium
Mamba	$O(L)$	1M	Low

- **Challenges**:
  - Weaker modeling ability for dynamic graph structures compared to GNNs
  - High parameter sensitivity, requiring careful tuning
  - Theoretical interpretability remains underdeveloped
- Use PyTorch's `SelectiveSSM` module for hardware-aware computation
- Apply channel-wise independence strategies when handling multivariate time series
- Combine with exponential moving average (EMA) to enhance training stability

Case study: In the task of electric load forecasting, the MambaTS model, enhanced with the variable scanning transformer (VST) mechanism, mean absolute error (MAE) reduced to 1.83, representing a 15% improvement over Transformer-based baselines [53,54].

5.1.13 PINN (physics-informed neural network)

Physics-informed neural networks (PINNs) are a class of neural architectures that embed physical laws—typically expressed as partial differential equations (PDEs)—directly into the training process. First introduced by Raissi, Perdikaris, and Karniadakis [55] in 2019, PINNs enable highly data-efficient learning by enforcing that networks satisfy governing equations, effectively regularizing them with domain knowledge, see Fig. 5.15.

The core idea of PINNs is to integrate physical knowledge directly into the loss function of neural networks, ensuring that the predicted outputs adhere to underlying physical laws:

1. **Physics-based loss function**: The training objective combines data loss and physics loss, typically defined as

$$\mathcal{L} = \mathcal{L}_{\text{data}} + \lambda_{\text{phys}} \mathcal{L}_{\text{phys}}, \tag{5.30}$$

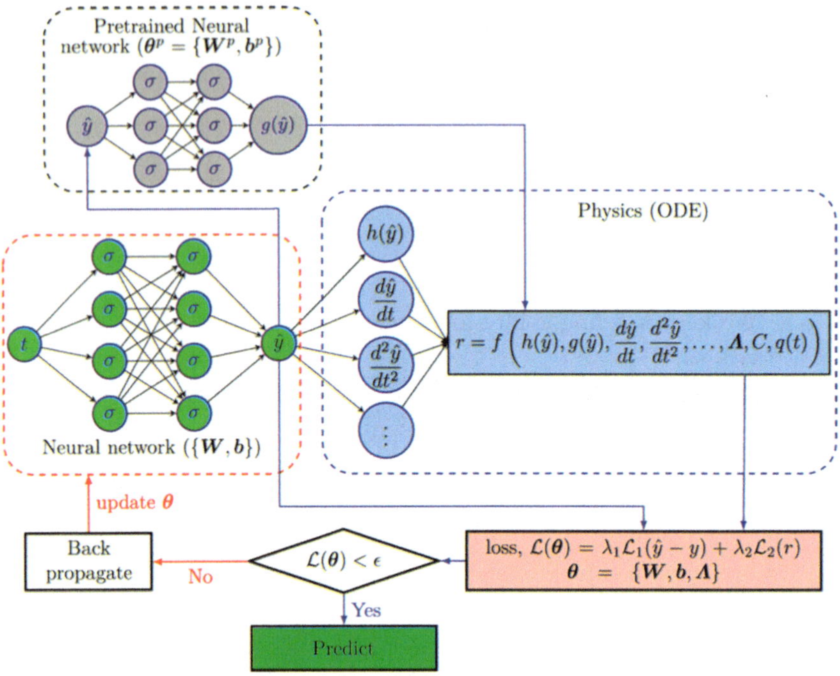

Figure 5.15 Structure of a physics-informed neural network (PINN).

where $\mathcal{L}_{data}$ measures the error between predictions and data, $\mathcal{L}_{phys}$ measures the residuals of the governing PDEs, and λ_{phys} is a balancing coefficient [55].

2. **Automatic differentiation**: PINNs leverage automatic differentiation (AD) to compute derivatives of the neural network outputs with respect to inputs, which are essential for evaluating PDE residuals [55].

- **Residual minimization** enforces the PDE constraints by minimizing the residual at randomly sampled collocation points:

$$\mathcal{L}_{phys} = \frac{1}{N_c}\sum_{i=1}^{N_c} |\mathcal{N}[u_\theta](\mathbf{x}_i)|^2, \tag{5.31}$$

where $\mathcal{N}$ denotes the differential operator, and u_θ is the neural network approximation [55].

- **Adaptive weighting strategies**: Techniques such as SoftAdapt and GradNorm dynamically adjust the loss weights to balance between data fitting and physics enforcement [56].

- **Domain decomposition**: Large or complex domains are partitioned into subdomains with separate PINNs trained per subdomain, enhancing scalability and convergence [57].
- **Forward problem solving**: Given initial and boundary conditions, PINNs predict the solution of PDEs such as Navier–Stokes, Schrödinger, or wave equations.
- **Inverse problem inference**: PINNs can infer unknown physical parameters or hidden fields from sparse measurements by embedding them as trainable variables [55,58].
- **Data assimilation**: Combining observational data and governing equations, PINNs improve the accuracy of prediction in data scarcity scenarios, see Table 5.2.

Table 5.2 Comparison between PINNs and traditional numerical methods.

Method	Data Requirement	Scalability	Physical Consistency
Finite element method (FEM)	Low	Moderate	High
Machine learning (without physics)	High	High	Low
PINN	Low-moderate	Moderate	High

- **Advantages**:
  - Incorporates physical knowledge without requiring labeled data [55]
  - Solves both forward and inverse problems seamlessly [58]
  - Flexible across various PDE types and boundary conditions
- **Challenges**:
  - Training can suffer from stiffness and convergence issues, especially in complex systems
  - Poor scalability to high-dimensional or multi-scale problems [58]
  - Balancing data and physics losses requires careful hyperparameter tuning [56]

- Employ gradient normalization or adaptive loss weighting to stabilize training [56]
- Use domain decomposition for complex geometries [57]
- Integrate transfer learning or pretraining strategies to accelerate convergence [59]

5.2 Probabilistic or stochastic models

5.2.1 Uncertainty quantification

Uncertainty quantification (UQ) plays a pivotal role in industrial soft sensing by providing probabilistic guarantees for model predictions [60,61]. In industrial processes where measurement noise and parameter drift are inevitable, UQ enables the characterization of both aleatoric (data-inherent) and epistemic (model-induced) uncertainties [62].

The predictive uncertainty can be decomposed as

$$\mathbb{V}(y|\mathbf{x}) = \underbrace{\mathbb{E}\theta[(\mu\theta(\mathbf{x}) - \mathbb{E}[\mu_\theta(\mathbf{x})])^2]}_{\text{Model uncertainty}} + \underbrace{\mathbb{E}\theta[\sigma_\theta^2(\mathbf{x})]}_{\text{Data uncertainty}} . \tag{5.32}$$

Confidence interval (CI) provides a range of values that is likely to contain the true value of an unknown population parameter with certain level of confidence [63]. It is widely used in statistics to quantify uncertainty in parameter estimation [64].

Given a sample statistic $\hat{\theta}$ (e.g., sample mean), a $100(1-\alpha)\%$ confidence interval for the population parameter θ is the interval

$$\left[\hat{\theta} - z_{\alpha/2} \cdot \sigma_{\hat{\theta}},\ \hat{\theta} + z_{\alpha/2} \cdot \sigma_{\hat{\theta}}\right], \tag{5.33}$$

where

- $\hat{\theta}$: point estimate of θ
- $\sigma_{\hat{\theta}}$: standard error of $\hat{\theta}$
- $z_{\alpha/2}$: the critical value from the standard normal distribution satisfying $\mathbb{P}(-z_{\alpha/2} \le Z \le z_{\alpha/2}) = 1 - \alpha$

Common example: Confidence interval for the mean: Assuming a sample $\{x_1, x_2, \ldots, x_n\}$ drawn from a normal distribution with known variance σ^2, the $100(1-\alpha)\%$ confidence interval for the population mean μ is

$$\left[\bar{x} - z_{\alpha/2} \cdot \frac{\sigma}{\sqrt{n}},\ \bar{x} + z_{\alpha/2} \cdot \frac{\sigma}{\sqrt{n}}\right]. \tag{5.34}$$

If σ is unknown and estimated from the sample, then the t-distribution is used instead:

$$\left[\bar{x} - t_{\alpha/2,n-1} \cdot \frac{s}{\sqrt{n}},\ \bar{x} + t_{\alpha/2,n-1} \cdot \frac{s}{\sqrt{n}}\right], \tag{5.35}$$

where s is the sample standard deviation, and $t_{\alpha/2,n-1}$ is the critical value from the Student's t-distribution with $n-1$ degrees of freedom.

Remark. A 95% confidence interval does **not** mean that there is a 95% probability that θ lies within the interval for a fixed dataset. It means that if we repeatedly draw samples and construct such intervals, approximately 95% of them will contain the true parameter θ.

- Widely used in hypothesis testing, model uncertainty quantification, and experimental analysis.
- Confidence intervals can be constructed for proportions, variances, regression coefficients, etc.

Industrial applications include:

- Chemical process monitoring with confidence intervals
- Predictive maintenance with failure probability estimation
- Energy consumption forecasting with uncertainty bounds

5.2.2 Quantile regression

Quantile regression extends traditional regression by estimating conditional quantiles of the response variable [65,66]. For industrial prediction tasks requiring asymmetric risk assessment, the quantile loss function is defined as

$$L_\tau(y, \hat{q}_\tau) = \begin{cases} \tau(y - \hat{q}_\tau) & \text{if } y \geq \hat{q}_\tau, \\ (1-\tau)(\hat{q}_\tau - y) & \text{otherwise.} \end{cases} \tag{5.36}$$

When $\tau = 0.5$, QR is a median regression. Key industrial implementations:

- Robust quality prediction in semiconductor manufacturing (90%/10% quantiles)
- Inventory management with quantile forecasts for supply chain optimization
- Safety-critical temperature prediction in nuclear reactors

5.2.3 Bayesian neural networks

Bayesian neural networks (BNNs) provide a principled framework for uncertainty-aware deep learning through posterior distributions over weights [67,68] (see Fig. 5.16):

$$p(\mathbf{w}|\mathcal{D}) = \frac{p(\mathcal{D}|\mathbf{w})p(\mathbf{w})}{p(\mathcal{D})}. \tag{5.37}$$

Practical implementation strategies for industry:

- Variational inference for real-time inference
- Monte Carlo dropout for uncertainty estimation

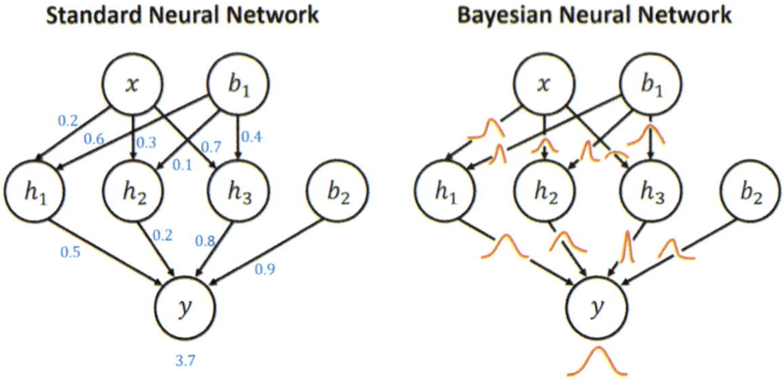

Figure 5.16 Structure of artificial neural network (ANN) and Bayesian neural network (BNN).

- Hamiltonian Monte Carlo for high-fidelity models

Industrial case studies demonstrate:

- 30% reduction in false alarms in petroleum refinery monitoring
- Improved generalization across different production batches
- Effective handling of sensor dropout scenarios

5.3 Interpretable and explainable models

5.3.1 KAN (Kolmogorov–Arnold networks)

In contrast to MLPs, which rely fundamentally on the universal approximation theorem, KANs derive their theoretical foundation from the Kolmogorov–Arnold representation theorem (KART) [69,70]. This seminal result establishes that any continuous multivariate function can be decomposed into finite compositions of univariate functions [71]. It is known that KART has catalyzed transformative advancements across dynamical systems, ergodic theory, and approximation theory [72]. By reimagining neural architectures through this lens, KANs depart from the rigid layered hierarchies of MLPs, instead adopting a mathematically grounded framework that dynamically adapts activation patterns through spline-parameterized univariate functions [73]. This paradigm shift not only preserves theoretical rigor but also enables flexible, interpretable modeling of complex functional relationships, a critical advantage for applications demanding both accuracy and transparency, such as climate forecasting [70].

The Kolmogorov–Arnold representation theorem states that any multivariate continuous function f, defined over a bounded domain and dependent on the vector $\mathbf{x} = [x_1, x_2, \ldots, x_n]$, can be expressed as a finite composition of continuous univariate functions. Specifically, any real-valued, smooth, and continuous function $f : [0, 1]^n \rightarrow \mathbb{R}$ admits the following representation:

$$f(\mathbf{x}) = \sum_{i=1}^{2n+1} \Phi_i \left(\sum_{j=1}^{n} \phi_{i,j}(x_j) \right), \tag{5.38}$$

where $\Phi_i : \mathbb{R} \rightarrow \mathbb{R}$ and $\phi_{i,j} : [0, 1] \rightarrow \mathbb{R}$ are referred to as the outer and inner functions, respectively. At first glance, this formulation appears to be highly promising for machine learning, as it reduces the challenge of modeling a high-dimensional function to the task of learning a polynomial number of univariate functions. A significant limitation arises from the fact that these univariate functions, however, can be highly non-smooth and irregular, making them difficult to approximate or learn in practice. Due to this challenge, the Kolmogorov–Arnold theorem has traditionally been overlooked in machine learning, viewed more as a theoretical curiosity than a practical tool. Recent advancements have reignited interest in this theorem, positioning it as a potential foundation for novel neural network architectures that explicitly draw on its constructive insights.

KANs employ dual nonlinear layers, structured with $2n+1$ intermediate terms. The KAN architecture is presented in Fig. 5.17. The core challenge lies in identifying suitable univariate functions $\phi_{i,j}$ and Φ_i to approximate the target function. For the one-dimensional $\phi_{i,j}$, *B-splines*, smooth curves defined by control points (knots), serve as an efficient approximation tool. These curves are characterized by their polynomial order k (typically $k = 3$) and the number of intervals G, which determine the segments between knots. During interpolation, $G + 1$ grid points ensure continuity and smoothness. While B-splines are emphasized here, alternative spline formulations could also be considered.

Eq. (5.38) maps to a two-layer architecture, where activation functions reside on edges rather than nodes, with the latter performing summation. However, this shallow configuration struggles to approximate arbitrary functions effectively using smooth splines. To address this limitation, recent work proposes generalized architectures with increased depth and width, enhancing the expressive power of KANs.

A KAN layer is defined by a matrix Φ of trainable univariate functions $\phi_{i,j}$, where $i = 1, \ldots, N_{\text{in}}$ and $j = 1, \ldots, N_{\text{out}}$ correspond to the input and output dimensions, respectively. As described earlier,

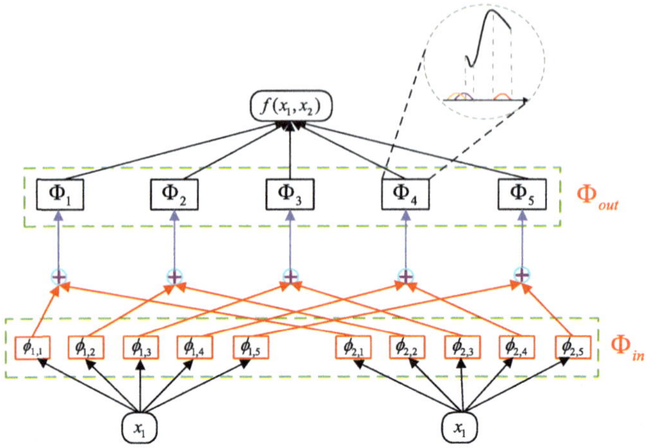

Figure 5.17 Illustration of the Kolmogorov–Arnold network (KAN).

each $\phi_{i,j}$ is parameterized via splines. Crucially, the Kolmogorov–Arnold representation theorem maps directly to a two-layer KAN. The inner layer operates on $N_{\text{in}} = n$ inputs and produces $N_{\text{out}} = 2n + 1$ outputs. The external layer transforms these $2n+1$ inputs into a single output ($N_{\text{out}} = 1$).

The topology of a KAN is defined by $[n_1, \ldots, n_{L+1}]$, where L denotes the network depth of the KAN. According to the Kolmogorov–Arnold theorem, KAN is shaped with dimension of $[n, 2n+1, 1]$. A generic deeper KAN can be expressed by the composition of L layers:

$$\mathbf{y} = \text{KAN}(\mathbf{x}) = (\mathbf{\Phi}_L \circ \mathbf{\Phi}_{L-1} \circ \ldots \circ \mathbf{\Phi}_1)\,\mathbf{x}. \tag{5.39}$$

It is worth noting that all operations in KANs are differentiable, enabling training via backpropagation. While rooted in mathematical elegance, KANs synergistically combine splines and MLPs, leveraging their complementary strengths. In KAN, splines excel at accurate low-dimensional approximation with multi-resolution flexibility but suffer from the curse of dimensionality due to poor compositional structure exploitation. MLPs mitigate dimensionality issues through feature learning but underperform splines in low-dimensional regimes due to suboptimal univariate function optimization. KANs inherently balance two levels of degrees of freedom. External freedom (architectural similarity to MLPs) enables compositional feature learning, whereas internal freedom (spline-parameterized edges) ensures precise univariate function approximation.

By adjusting the number of layers L or grid dimension G, the model complexity scales with added parameters. This parametric flexibility positions KANs as a viable alternative to conventional deep learning models reliant on rigid MLP architectures, motivating their extended exploration in this work.

5.4 Hybrid models

Based on the aforementioned analysis, single models exhibit inherent limitations when applied to complex time series or soft sensing problems. By strategically combining the strengths of the aforementioned individual models, it becomes possible to capitalize on their complementary advantages while mitigating their respective weaknesses. Consequently, hybrid modeling approaches have been developed and widely adopted to achieve superior predictive performance and robustness [74].

A particularly effective and conceptually straightforward methodology is the **weighted hybrid model**. This approach leverages multiple distinct time series or regression models (e.g., ARIMA, exponential smoothing, support vector regression, neural networks, Gaussian process regression) to generate independent predictions for the target variable. Subsequently, a final prediction $\hat{y}_t$ is obtained by computing a weighted average of these individual model forecasts:

$$\hat{y}_t = \sum_{i=1}^{N} w_{i,t} \cdot \hat{y}_{i,t}, \tag{5.40}$$

where:

- $\hat{y}_{i,t}$ is the prediction from the ith base model at time t,
- $w_{i,t}$ is the weight assigned to the ith model at time t,
- N is the total number of base models combined.

The weights ($w_{i,t}$) are critical and typically satisfy $\sum_{i=1}^{N} w_{i,t} = 1$ and $w_{i,t} \geq 0$. These weights can be:

1. **Static**, determined offline based on historical performance metrics (e.g., minimizing validation error) and remain constant during operation.
2. **Dynamic**, adapted online over time, often reflecting the recent predictive accuracy or reliability of each constituent model. For instance, models performing better on recent data points may receive higher weights.

Advantages for time series and soft sensing

- **Enhanced accuracy:** Combines diverse modeling perspectives, often leading to predictions that outperform any single base model.
- **Increased robustness:** Reduces the risk associated with relying on a single model structure, making the prediction system less sensitive to model-specific assumptions or anomalies in the data. This is crucial for soft sensors where process conditions can vary.
- **Flexibility:** Can incorporate models suited for capturing different aspects of the data (e.g., linear trends, seasonality, non-linear dynamics, uncertainty).
- **Mitigation of overfitting:** The averaging effect can help smooth out idiosyncratic errors from individual models.

Weighted hybrid models are thus a powerful tool for improving the reliability and accuracy of predictions in demanding applications like forecasting complex time series and developing robust soft sensors for industrial processes.

5.5 Conclusions

This chapter has presented a comprehensive examination of both deterministic and statistical machine learning models for industrial applications. Key contributions include:

- A systematic analysis of diverse modeling approaches, highlighting their respective advantages and inherent limitations
- Practical insights into model deployment for time-series prediction and soft sensing applications in industrial settings
- Introduction of the weighted hybrid modeling paradigm as a robust framework for enhancing prediction accuracy and reliability

The comparative evaluation reveals that although individual models exhibit specific strengths, their limitations in handling complex industrial dynamics necessitate advanced solutions. The hybrid weighted approach emerges as a particularly promising methodology, combining complementary model capabilities to overcome singular weaknesses. This synergistic framework demonstrates significant potential for improving decision support systems in process industries through enhanced predictive performance and operational robustness.

References

[1] Adebiyi A. Ariyo, Adewumi O. Adewumi, Charles K. Ayo, Stock price prediction using the ARIMA model, in: 2014 UKSim-AMSS 16th International Conference on Computer Modelling and Simulation, IEEE, 2014, pp. 106–112.
[2] Javier Contreras, Rosario Espinola, Francisco J. Nogales, Antonio J. Conejo, ARIMA models to predict next-day electricity prices, IEEE Transactions on Power Systems 18 (3) (2003) 1014–1020.
[3] Robert H. Shumway, David S. Stoffer, ARIMA models, in: Time Series Analysis and Its Applications: With R Examples, Springer, 2017, pp. 75–163.
[4] Biao Zhang, Deyi Xiong, Jinsong Su, Neural machine translation with deep attention, IEEE Transactions on Pattern Analysis and Machine Intelligence 42 (1) (2018) 154–163.
[5] S. Agatonovic-Kustrin, Rosemary Beresford, Basic concepts of artificial neural network (ANN) modeling and its application in pharmaceutical research, Journal of Pharmaceutical and Biomedical Analysis 22 (5) (2000) 717–727.
[6] James W. Taylor, A quantile regression neural network approach to estimating the conditional density of multiperiod returns, Journal of Forecasting 19 (4) (2000) 299–311.
[7] David Salinas, Valentin Flunkert, Jan Gasthaus, Tim Januschowski, DeepAR: probabilistic forecasting with autoregressive recurrent networks, International Journal of Forecasting 36 (3) (2020) 1181–1191.
[8] J. Geng, C. Yang, Y. Li, et al., MPA-RNN: a novel attention-based recurrent neural networks for total nitrogen prediction, IEEE Transactions on Industrial Informatics 18 (10) (2022) 6516–6525.
[9] Larry Medsker, Lakhmi C. Jain, Recurrent Neural Networks: Design and Applications, CRC Press, 1999.
[10] Jeffrey L. Elman, Finding structure in time, Cognitive Science 14 (2) (1990) 179–211.
[11] Michael I. Jordan, Serial order: a parallel distributed processing approach, in: Advances in Psychology, vol. 121, Elsevier, 1997, pp. 471–495.
[12] Rohaifa Khaldi, Abdellatif El Afia, Raddouane Chiheb, Siham Tabik, What is the best RNN-cell structure to forecast each time series behavior?, Expert Systems with Applications 215 (2023) 119140.
[13] Paul J. Werbos, Backpropagation through time: what it does and how to do it, Proceedings of the IEEE 78 (10) (2002) 1550–1560.
[14] George Bird, Maxim E. Polivoda, Backpropagation through time for networks with long-term dependencies, arXiv preprint, arXiv:2103.15589, 2021.
[15] Mahrad Pisheh Var, Minimalistic Adaptive Dynamic-Programming Agents for Memory-Driven Exploration, PhD thesis, University of Essex, 2024.
[16] Sepp Hochreiter, Jürgen Schmidhuber, Long short-term memory, Neural Computation 9 (8) (1997) 1735–1780.
[17] Kyunghyun Cho, Bart Van Merriënboer, Caglar Gulcehre, Dzmitry Bahdanau, Fethi Bougares, Holger Schwenk, Yoshua Bengio, Learning phrase representations using RNN encoder-decoder for statistical machine translation, arXiv preprint, arXiv: 1406.1078, 2014.
[18] Mike Schuster, Kuldip K. Paliwal, Bidirectional recurrent neural networks, IEEE Transactions on Signal Processing 45 (11) (1997) 2673–2681.
[19] Ilya Sutskever, Oriol Vinyals, Quoc V. Le, Sequence to sequence learning with neural networks, Advances in Neural Information Processing Systems 27 (2014).
[20] Ahmed Tealab, Time series forecasting using artificial neural networks methodologies: a systematic review, Future Computing and Informatics Journal 3 (2) (2018) 334–340.
[21] Alex Graves, Generating sequences with recurrent neural networks, arXiv preprint, arXiv:1308.0850, 2013.

[22] Subhashini Venugopalan, Marcus Rohrbach, Jeffrey Donahue, Raymond Mooney, Trevor Darrell, Kate Saenko, Sequence to sequence-video to text, in: Proceedings of the IEEE International Conference on Computer Vision, 2015, pp. 4534–4542.
[23] Max Jaderberg, Karen Simonyan, Andrew Zisserman, et al., Spatial transformer networks, Advances in Neural Information Processing Systems 28 (2015).
[24] Chu Zhang, Huixin Ma, Lei Hua, Wei Sun, Muhammad Shahzad Nazir, Tian Peng, An evolutionary deep learning model based on TVFEMD, improved sine cosine algorithm, CNN and BiLSTM for wind speed prediction, Energy 254 (2022) 124250.
[25] Wenshuai Lin, Bin Zhang, Hongyi Li, Renquan Lu, Multi-step prediction of photovoltaic power based on two-stage decomposition and BILSTM, Neurocomputing 504 (2022) 56–67.
[26] Ronil Barua, Anil K. Sharma, Dynamic Black Litterman portfolios with views derived via CNN-BiLSTM predictions, Finance Research Letters 49 (2022) 103111.
[27] Jin Chen, Xingchen Li, Ye Xiao, Hao Chen, Yong Zhao, FRA-LSTM: a vessel trajectory prediction method based on fusion of the forward and reverse sub-network, arXiv preprint, arXiv:2201.07606, 2022.
[28] Chaoshun Li, Geng Tang, Xiaoming Xue, Adnan Saeed, Xin Hu, Short-term wind speed interval prediction based on ensemble GRU model, IEEE Transactions on Sustainable Energy 11 (3) (2019) 1370–1380.
[29] Farhad Mortezapour Shiri, Thinagaran Perumal, Norwati Mustapha, Raihani Mohamed, A comprehensive overview and comparative analysis on deep learning models: CNN, RNN, LSTM, GRU, arXiv preprint, arXiv:2305.17473, 2023.
[30] Jun Han, Claudio Moraga, The influence of the sigmoid function parameters on the speed of backpropagation learning, in: International Workshop on Artificial Neural Networks, Springer, 1995, pp. 195–201.
[31] Razvan Pascanu, Tomas Mikolov, Yoshua Bengio, On the difficulty of training recurrent neural networks, ICML (2013).
[32] Rafal Jozefowicz, Wojciech Zaremba, Ilya Sutskever, An empirical exploration of recurrent network architectures, in: ICML, 2015.
[33] Junyoung Chung, Caglar Gulcehre, Kyunghyun Cho, Yoshua Bengio, Empirical evaluation of gated recurrent neural networks on sequence modeling, arXiv preprint, arXiv:1412.3555, 2014.
[34] Kyunghyun Cho, et al., Learning phrase representations using RNN encoder–decoder for statistical machine translation, in: EMNLP, 2014.
[35] Dzmitry Bahdanau, Kyunghyun Cho, Yoshua Bengio, Neural machine translation by jointly learning to align and translate, in: ICLR, 2015.
[36] Alex Graves, Abdel-rahman Mohamed, Geoffrey E. Hinton, Speech recognition with deep recurrent neural networks, in: ICASSP, 2013.
[37] Hansika Hewamalage, Christoph Bergmeir, Kasun Bandara, Recurrent neural networks for time series forecasting: current status and future directions, International Journal of Forecasting (2020).
[38] Jiuxiang Gu, Zhenhua Wang, Jason Kuen, Lianyang Ma, Amir Shahroudy, Bing Shuai, Ting Liu, Xingxing Wang, Gang Wang, Jianfei Cai, et al., Recent advances in convolutional neural networks, Pattern Recognition 77 (2018) 354–377.
[39] Hamed Habibi Aghdam, Elnaz Jahani Heravi, et al., Guide to Convolutional Neural Networks, Springer, New York, NY, 2017, 10(978-973) 51.
[40] Xianxu Hou, Ke Sun, Linlin Shen, Guoping Qiu, Improving variational autoencoder with deep feature consistent and generative adversarial training, Neurocomputing 341 (2019) 183–194.
[41] Qingsong Wen, Tian Zhou, Chaoli Zhang, Weiqi Chen, Ziqing Ma, Junchi Yan, Liang Sun, Transformers in time series: a survey, arXiv preprint, arXiv:2202.07125, 2022.

[42] Bryan Lim, Sercan O. Arik, Nicolas Loeff, Tomas Pfister, Temporal fusion transformers for interpretable multi-horizon time series forecasting, in: NeurIPS, 2019.
[43] Shreshth Tuli, Giuliano Casale, Nicholas R. Jennings, TranAD: deep transformer networks for anomaly detection in multivariate time series data, arXiv preprint, arXiv:2201.07284, 2022.
[44] X. Xu, et al., Anomaly transformer: time series anomaly detection with association discrepancy, in: ICLR, 2022.
[45] Shaojie Bai, J. Zico Kolter, Vladlen Koltun, An empirical evaluation of generic convolutional and recurrent networks for sequence modeling, arXiv preprint, arXiv: 1803.01271, 2018.
[46] Thomas N. Kipf, Max Welling, Semi-supervised classification with graph convolutional networks, arXiv preprint, arXiv:1609.02907, 2016.
[47] Petar Veličković, Guillem Cucurull, Arantxa Casanova, Adriana Romero, Pietro Liò, Yoshua Bengio, Graph attention networks, arXiv preprint, arXiv:1710.10903, 2017.
[48] William L. Hamilton, Rex Ying, Jure Leskovec, Inductive representation learning on large graphs, in: Advances in Neural Information Processing Systems (NeurIPS), 2017.
[49] Emanuele Rossi, Ben Chamberlain, Fabrizio Frasca, Davide Eynard, Federico Monti, Michael Bronstein, Temporal graph networks for deep learning on dynamic graphs, in: ICML Workshop on Graph Representation Learning, 2020.
[50] Albert Gu, Tri Dao, Mamba: linear-time sequence modeling with selective state spaces, arXiv preprint, arXiv:2312.00752, 2023.
[51] Tri Dao, Albert Gu, Transformers are ssms: generalized models and efficient algorithms through structured state space duality, arXiv preprint, arXiv:2405.21060, 2024.
[52] Weihao Yu, Xinchao Wang, Mambaout: do we really need Mamba for vision?, in: Proceedings of the Computer Vision and Pattern Recognition Conference, 2025, pp. 4484–4496.
[53] Xiuding Cai, Yaoyao Zhu, Xueyao Wang, Yu Yao, MambaTS: improved selective state space models for long-term time series forecasting, arXiv preprint, arXiv:2405.16440, 2024.
[54] Jun Li, Haoxing Zhou, Short-term load forecasting under extreme weather based on Mamba–Transformer model, in: International Joint Conference on Energy, Electrical and Power Engineering, Springer, 2024, pp. 97–107.
[55] Maziar Raissi, Paris Perdikaris, George E. Karniadakis, Physics-informed neural networks: a deep learning framework for solving forward and inverse problems involving nonlinear partial differential equations, Journal of Computational Physics 378 (2019) 686–707, https://doi.org/10.1016/j.jcp.2018.10.045.
[56] Levi McClenny, Ulisses Braga-Neto, Self-adaptive physics-informed neural networks using a soft attention mechanism, arXiv preprint, arXiv:2009.04544, 2020.
[57] Ameya D. Jagtap, George E. Karniadakis, Extended physics-informed neural networks (XPINNs): a generalized space-time domain decomposition based deep learning framework for nonlinear partial differential equations, Communications in Computational Physics 28 (5) (2020) 2002–2041, https://doi.org/10.4208/cicp.OA-2020-0164.
[58] George Em Karniadakis, Ioannis G. Kevrekidis, Lu Lu, Paris Perdikaris, Sifan Wang, Liu Yang, Physics-informed machine learning, Nature Reviews Physics 3 (6) (2021) 422–440.
[59] Mahyar Jahani-Nasab, Mohamad Ali Bijarchi, Enhancing convergence speed with feature enforcing physics-informed neural networks using boundary conditions as prior knowledge, Scientific Reports 14 (1) (2024) 23836.
[60] Bochun Yue, Kai Wang, Hongqiu Zhu, Chunhua Yang, A domain-knowledge embedded framework for soft sensing in complex industrial processes with cascading equipment, IEEE Transactions on Industrial Informatics 21 (2025) 2510–2519, https://doi.org/10.1109/TII.2024.3507937.

[61] Minjung Lee, Jinsoo Bae, S. Kim, Uncertainty-aware soft sensor using Bayesian recurrent neural networks, Advanced Engineering Informatics 50 (2021) 101434, https://doi.org/10.1016/j.aei.2021.101434.
[62] Xiaoning Zhang, Fang Fang, Jiaqi Wang, Probabilistic solar irradiation forecasting based on variational Bayesian inference with secure federated learning, IEEE Transactions on Industrial Informatics 17 (11) (2020) 7849–7859.
[63] D.P. Turner, H. Deng, T. Houle, Understanding and applying confidence intervals, Headache: The Journal of Head and Face Pain 60 (2020), https://doi.org/10.1111/head.13991.
[64] L. Gardes, Samuel Maistre, Nonparametric asymptotic confidence intervals for extreme quantiles, Scandinavian Journal of Statistics 50 (2022) 825–841, https://doi.org/10.1111/sjos.12610.
[65] Vilde Jensen, Filippo Maria Bianchi, Stian Normann Anfinsen, Ensemble conformalized quantile regression for probabilistic time series forecasting, IEEE Transactions on Neural Networks and Learning Systems 35 (7) (2022) 9014–9025.
[66] Jianming Hu, Jingwei Tang, Zhi Liu, A novel time series probabilistic prediction approach based on the monotone quantile regression neural network, Information Sciences 654 (2024) 119844.
[67] Faming Liang, Bayesian neural networks for nonlinear time series forecasting, Statistics and Computing 15 (1) (2005) 13–29.
[68] Ozan Kocadağlı, Barış Aşıkgil, Nonlinear time series forecasting with Bayesian neural networks, Expert Systems with Applications 41 (15) (2014) 6596–6610.
[69] Alexander Dylan Bodner, Antonio Santiago Tepsich, Jack Natan Spolski, Santiago Pourteau, Convolutional Kolmogorov–Arnold networks, arXiv preprint, arXiv:2406.13155, 2024.
[70] Vikas Dhiman Kan, Kolmogorov–Arnold networks: a review. Retrieved from, 2024.
[71] Ziming Liu, Yixuan Wang, Sachin Vaidya, Fabian Ruehle, James Halverson, Marin Soljačić, Thomas Y. Hou, Max Tegmark, KAN: Kolmogorov–Arnold networks, arXiv preprint, arXiv:2404.19756, 2024.
[72] Cristian J. Vaca-Rubio, Luis Blanco, Roberto Pereira, Màrius Caus, Kolmogorov–Arnold networks (KANs) for time series analysis, arXiv preprint, arXiv:2405.08790, 2024.
[73] Ag Stephens, Philip James, David Alderson, Stephen Pascoe, Simon Abele, Alan Iwi, Peter Chiu, The challenges of developing an open source, standards-based technology stack to deliver the latest UK climate projections, International Journal of Digital Earth 5 (1) (2012) 43–62.
[74] Yun Wang, Houhua Xu, Runmin Zou, Fan Zhang, Qinghua Hu, Dynamic non-constraint ensemble model for probabilistic wind power and wind speed forecasting, Renewable and Sustainable Energy Reviews 204 (2024) 114781.

PART 3

Applications

CHAPTER 6

A novel semi-supervised soft sensor framework based on convolutional Laplacian ELM: CNN-LapsELM

6.1 Introduction

In the electrolytic aluminum industry, the difference between the temperatures of the electrolyte and initial crystal is called a superheat degree (SD) [1–4]. The SD is one of the important indexes used to evaluate the performance of the electrolytic cell, which can reflect the current work situation of the electrolytic cell, such as the amount of aluminum. However, how to accurately measure the current electrolytic cell's SD is an unresolved challenge. Manual measurement is a commonly used method, which can be affected by many aspects, such as the precision of the measuring instrument, the error of the manual reading, etc. [5]. Electrolytic aluminum industry site is in an environment of high temperature, high humidity, and high concentration of corrosive gas and dust. Due to the harsh production environment, the traditional manual measurement of the temperature of the initial crystal will be affected. At the same time the accuracy of the initial crystal temperature, the quality of the measuring instrument, and the accuracy of the manual reading are directly related to the SD results [6]. To sum up, in the actual process of electrolytic aluminum, the temperature of the initial crystal is difficult to measure, and the cost of off-line analysis measurement is large.

Recently, many network-based methods have been proposed for SD identification. For example, Zeng et al. [7] analyzed the effect of superheat on the aluminum industrial process and provided an effective control technique of SD. Chen et al. [8] used the semantic network based on intuitionistic fuzzy directed hyper-graphs to aluminum electrolysis cell condition identification. This method proposed a new insight for the SD identification in the aluminum production process. Liu et al. [9] put forward the method by using random forest algorithm to predict the SD, which uses

Learning-Based Soft Sensing and Predictions for Process Industries
https://doi.org/10.1016/B978-0-44-336759-5.00015-9

historical data to calculate the SD, and then divided it into high and low categories; however, this method was greatly affected by data noise, and the prediction efficiency was not satisfactory.

Convolutional neural network (CNN) is a popular network architecture, which has been a research point applied to various fields. Zuo et al. [10] improved the existing convolutional neural network by combining the visual recognition algorithm and the deep learning theory and solved the problems such as over-learning of the mapping function and insufficient convergence of the loss function in the super-resolution image restoration algorithm. Their works effectively improved the image definition, visual effect, and objective parameter evaluation index. Zhao et al. [11] aimed to solve the problem of low accuracy and time-consuming traditional computer-aided diagnostics; they proposed a new method of breast disease diagnosis based on convolutional neural networks multi-layer feature fusion and extreme learning machine. This method significantly improved the accuracy of breast disease diagnosis and provided insight to develop our method.

Due to the result of the labeled training, images are scarce in the aluminum electrolysis industry, so labeling these flame hole images manually costs a lot of material and human resource. Based on the above analysis, we propose an SD classification method that combines the CNN and semi-supervised extreme learning machine. Firstly, we use a convolutional neural network to extract deep features from flame hole images. Secondly, we take the fused features extracted from the flame hole image as the input of the Laplacian extreme learning machine to train the network model until a satisfactory recognition accuracy is achieved. Finally, we use the trained network model to predict the SD of the electrolytic cell. The main advantages of CNN-LapsELM are as follows:

- Both labeled data and unlabeled images are used to reduce the cost of labeling the flame hole image artificially.
- Making full use of the efficiency of CNN feature extraction and the comprehensiveness of fusion, as well as the high computation efficiency of the ELM training process.
- Laplacian regularization is used in CNN-LapsELM for local geometry information extraction, which fully excavates information contained in all available data.
- The fusion of original calculated features for top-level ELM training process can greatly improve the accuracy for SD identification, and the experiment result indicates the SD accuracy is as high as 87%.

6.2 Problem statement

In the SD soft sensing, the main task is using the interrelated variables to estimate the hard-to-measure SD. Different from the past methods, some factors are prominent and must be considered:

- The apparent features are calculated for building fused features in the CNN-LapsELM method, with the fusion feature furthest from the expert experience and other factors that can fully elevate the SD identification performance.
- The labeled frame hole samples are scant in the industrial aluminum production process, the information contained in a large number of unlabeled data are discarded, and manifold regularization for extracting the information contained in both labeled and unlabeled data can further save cost while the accuracy rate is ensured.
- Only judging the tank condition by experimented workers for SD identification is limited and objective; however, due to the cognitive limitations and inconsistencies of experts and operators, the SD identification by workers is objective and sometimes inaccurate. An integrated method based on CNN for deep extraction and calculated features for fusion is proposed to identify the SD situation.

Given a labeled training data set $\{\mathbf{x}_i, \mathbf{y}_i\}_{i=1}^{l}$ and unlabeled data $\{\mathbf{x}_j\}_{j=1}^{u}$, the Laplacian-based semi-supervised methods can fully excavate the geometry distribution information contained in all available data. Semi-supervised learning framework is built on the following two assumptions:

- Both labeled data $\mathbf{X}_l$ and unlabeled data $\mathbf{X}_u$ are drawn from the same marginal distribution P_X.
- The conditional probabilities $P(\mathbf{y}|\mathbf{x}_1)$ and $P(\mathbf{y}|\mathbf{x}_2)$ are similar if $\mathbf{x}_1$ and $\mathbf{x}_2$ are close in the feature space.

The latter assumption (smoothness assumption) is widely applied in machine learning domains. The regularization function derived from this assumption is

$$L_{\mathrm{m}} = \frac{1}{2}\sum_{i,j} u_{ij} \left\| P(\mathbf{y}|\mathbf{x}_i) - P(\mathbf{y}|\mathbf{x}_j) \right\|^2, \tag{6.1}$$

where u_{ij} measures pairwise similarity between samples $\mathbf{x}_i$ and $\mathbf{x}_j$, typically computed using the Gaussian kernel:

$$u_{ij} = \exp\left(-\frac{\|\mathbf{x}_i - \mathbf{x}_j\|^2}{2\sigma^2}\right). \tag{6.2}$$

Eq. (6.1) can be expressed in matrix form:

$$\hat{L}_m = \mathrm{tr}(\hat{\mathbf{Y}}^\top \mathbf{L}\hat{\mathbf{Y}}), \tag{6.3}$$

where $\mathrm{tr}(\cdot)$ denotes the matrix trace, $\mathbf{L} = \mathbf{D} - \mathbf{U}$ is the graph Laplacian, $\mathbf{U} = [u_{ij}]$ is the similarity matrix, and $\mathbf{D}$ is the diagonal degree matrix with elements

$$D_{ii} = \sum_{j=1}^{l+u} u_{ij}. \tag{6.4}$$

Based on the above analysis and insights, we proposed the CNN-LapsELM algorithm for SD identification, and the details are given in the following section.

6.3 Methodology

The core idea is to fuse the different features of flame images. Therefore the first stage is compute or extract features from the flame hole images. All the fusing features are given in Table 6.1. Specifically, the detailed calculation and description of these features are given in Tables 6.2, 6.3, and 6.4.

Table 6.1 Types of features extracted from flame hole images.

No.	Tags	Feature descriptions
1	Deep features	256 deep features extracted from CNN
2	Texture features	Flame hole moment, entropy, energy
3	Color features	Color mean, color variance, color skew, color peak
4	Gray features	Change ratio of corner point, change ratio of average gray, correlation coefficient, average gray value

Table 6.2 Texture features description of flame hole images.

Features	Calculation equations	Parameter descriptions
Flame hole moment	$m_1 = \sum_{i,j} \frac{Cov(i,j)}{1+(i-j)^2}$	$Cov(i,j)$: gray covariance matrix element at position (i,j)
Entropy	$m_2 = \sum_{i,j} Cov(i,j) \cdot \left(-\ln Cov(i,j)\right)$	
Energy	$m_3 = \sum_{i,j} [Cov(i,j)]^2$	

Table 6.3 Description of color features of flame hole images.

Features	Calculation equations	Parameter descriptions
Color mean	$s_1 = \sum_{i=0}^{N} d_i p(d_i)$	N: gray value of the flame hole image
Color variance	$s_2 = \sum_{i=0}^{N-1} (d_i - n)^2 p(d_i)$	$p(d_i)$: the ratio of pixel with gray value equals to i
Color skew	$s_3 = \frac{1}{s_1^2} \sum_{i=0}^{N-1} (d_i - n)^3 p(d_i)$	
Color peak	$s_4 = \frac{1}{s_1^2} \sum_{i=0}^{N-1} (d_i - n)^4 p(d_i)$	

Table 6.4 Description of gray features of flame hole images.

Features	Calculation equations	Parameter descriptions
Change ratio of corner point	$t_1 = \frac{V_i - V_j}{J - i}$	V_i is the number of corner point in the ith moment.
Change ratio of average gray	$t_2 = \frac{S_i - S_j}{J - i}$	S_i denotes the mean gray value of ith seconds.
Correlation coefficient	$t_3 = \frac{\sum_k \sum_h (k - \mu_{x_k})(h - \mu_{y_k}) p(k, h)}{\sigma_x \sigma_y}$	μ_{x_k}, μ_{y_k} and σ_x, σ_y denote the mean values and variances of $p(k, h)$.
Average gray value	$t_4 = \frac{\sum_{i=1}^{N} \sum_{j=1}^{M} f(i,j)}{N \times M}$	$f(i,j)$ represents the gray value at pixel (i,j), and N and M are the numbers of rows and columns of flame hole images, respectively.

Then the fusion feature matrix is given as

$$\mathbf{M}_{\text{fusion}} = [\mathbf{M}_{\text{deep}}, \mathbf{M}_{\text{texture}}, \mathbf{M}_{\text{color}}, \mathbf{M}_{\text{gray}}], \tag{6.5}$$

where $\mathbf{M}_{\text{deep}}$ denotes the deep feature extracted from CNN, $\mathbf{M}_{\text{color}}$ is the color feature vector, $\mathbf{M}_{\text{texture}}$ is the texture characteristics, and $\mathbf{M}_{\text{gray}}$ is the feature input variable from gray flame hole images.

CNN-LapsELM flowchart is presented in Fig. 6.1. In the proposed framework, CNN is used to extract the deep features from the original flame hole images. A six-layer network with two convolution layers, two

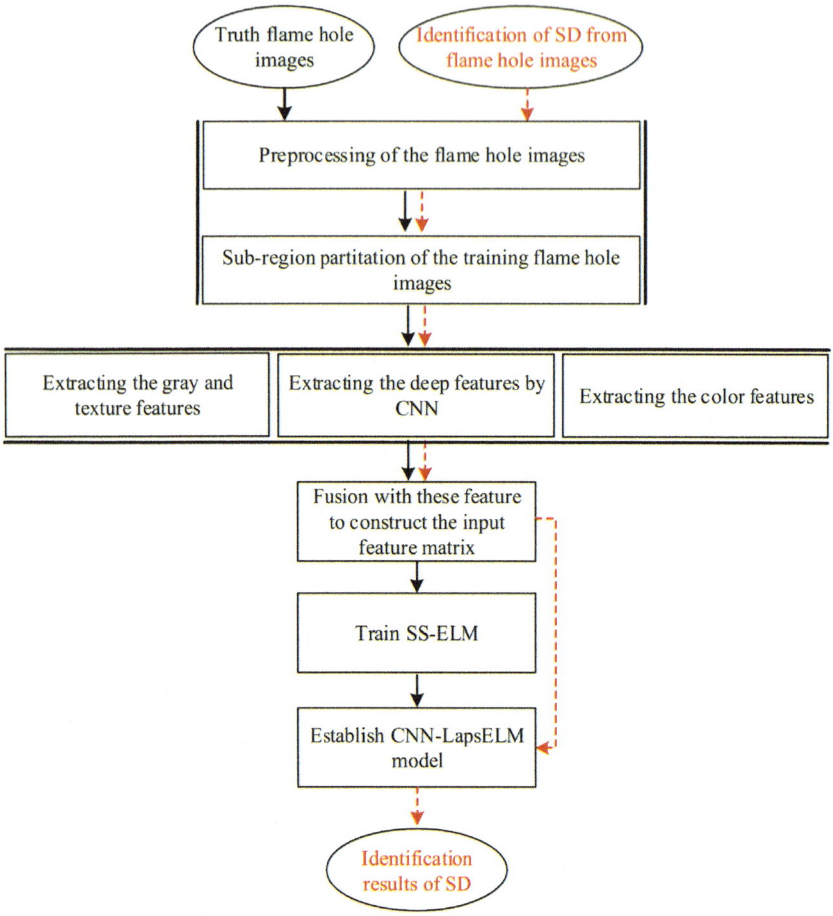

Figure 6.1 The flowchart of CNN-LapsELM.

max-pooling layers, and two fully connected layers is given. The input of CNN is the 48 × 48-dimensional flame hole images. Finally, a 256-dimensional deep feature matrix is extracted. The model training procedure is introduced with Laplacian regulation. The diagonal matrix D is an augmented matrix of dimension $l + u$, which covers the information of l labeled images and u unlabeled samples. By substituting the Laplacian regularization constraints into the objective function, the loss function is expressed as

$$\Gamma_{\text{LapsELM}} = \min_{\boldsymbol{\beta} \in R^{n_h \times n_0}} \frac{1}{2}\|\boldsymbol{\beta}\|^2 + \frac{C}{2}\|\mathbf{Y} - \mathbf{J}\boldsymbol{\beta}\|^2 + \frac{\lambda}{2}\operatorname{Tr}\left(\boldsymbol{\beta}^T\mathbf{J}^T\mathbf{L}\beta\right), \tag{6.6}$$

where $\mathbf{Y} \in \mathbb{R}^{(l+u)\times n_o}$ is the augmented training target with first l rows equal to $\mathbf{Y}_l$ and the rest rows set to 0, β is the hidden weight that needs to fine-tune, Γ_{LapsELM} defines the total loss function, and $\mathbf{C}$ is the $(l+u) \times (l+u)$ diagonal matrix with first l diagonal elements $[\mathbf{C}]_{ii} = C_i$ and the remaining equal to 0. Notably, this augmented training matrix introduces all the training targets that can learn the available images. By setting the gradient of the above equation to 0, the output weights of CNN-LapsELM can be calculated:

$$\boldsymbol{\beta}^* = \left(\mathbf{I}_{n_h} + \mathbf{J}^T\mathbf{C}\mathbf{J} + \lambda\mathbf{J}^T\mathbf{L}\mathbf{J}^{-1}\mathbf{J}^T\right)\mathbf{C}\mathbf{Y}, \tag{6.7}$$

where $\mathbf{I}_{n_h}$ is the $n_h \times n_h$ identity matrix. When the number of labeled data is less than the hidden neurons, the alternative solution of CNN-LapsELM is

$$\boldsymbol{\beta}^* = \mathbf{J}^T\left(\mathbf{I}_{l+u} + \mathbf{C}\mathbf{J}^T + \lambda\mathbf{L}\mathbf{J}\mathbf{J}^T\right)^{-1}\mathbf{C}\mathbf{Y}, \tag{6.8}$$

where $\mathbf{I}_{(l+u)}$ is the $(l+u) \times (l+u)$ identity matrix, and $\mathbf{Y}$ is the augmented matrix whose first l rows are equal to those of $\mathbf{J}$ and the last u rows are set to 0. If we set $\lambda = 0$ in (6.7) and (6.8), then CNN-LapsELM reduces to a conventional ELM.

6.4 Analysis of results

To evaluate the performances of our proposed CNN-LapsELM, two supervised algorithms, SVM [12] and ELM [13], are selected as the baseline classifiers. Some other state-of-the-art algorithms, DBN [14], CNN [15], SS-ELM [16], SAE, and CNN-ELM [17], are also compared. We see that with only a few number of labeled data points available, the performance of CNN- LapsELM greatly outperforms SVM and ELM. This reveals that the performance of the proposed algorithm is better than a traditional Laplacian regularization. Accuracy, training time and testing time were used to evaluate the performances of these algorithms. The RMSE index is used to evaluate the robustness of all the competitive algorithms. The results show that the proposed CNN-LapsELM has better performance than SVM and ELM in all training images, which also indicates that our proposed algorithm is able to fully leverage the unlabeled data to enhance the performances when compared to other supervised algorithms. The results present the testing time for CNN-LapsELM, SVM, and ELM. They show that the testing performance is better than those of CNN, ELM, SAE, and SS-ELM due to the Laplacian regularization, which can fully excavate the geometry information of sample distribution. However, the training time is relatively not improved, because the computation of Laplacian regularization term with all the available images is time-consuming. Fig. 6.2 shows the performance of CNN-LapsELM, SS-ELM,

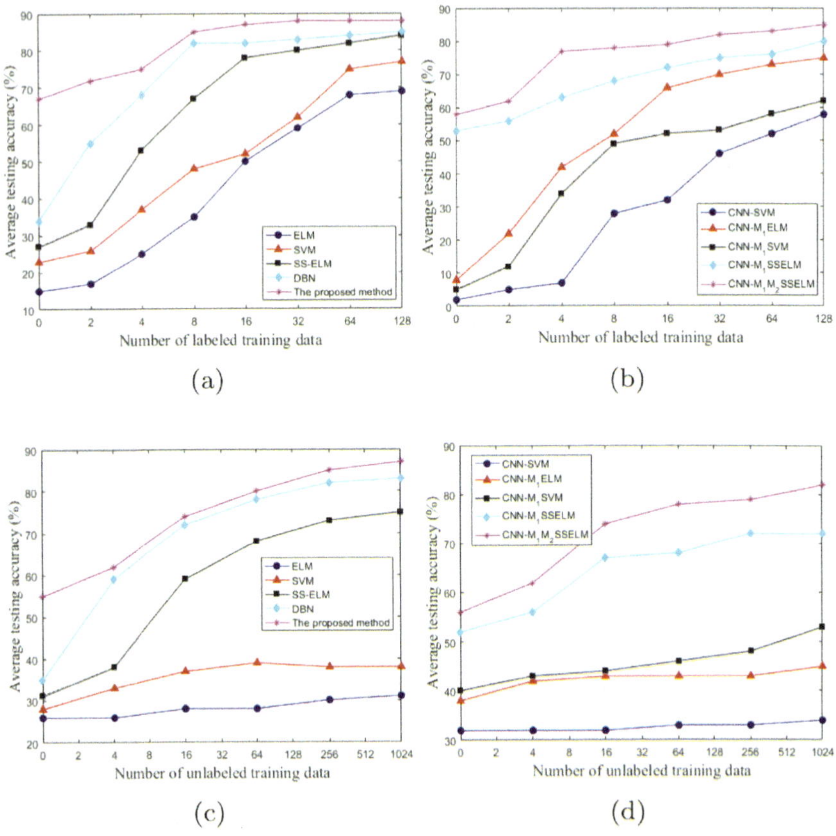

Figure 6.2 Performance comparison with various labeled and unlabeled flame hole images. Notably, the we gradually increase the number of labeled data and different features for ELM.

and ELM with various proportions of unlabeled data. We continuously added the number of the unlabeled set, while the labeled set, test set, and validation set remained unchanged. It demonstrates that the test error drops significantly when more unlabeled patterns are added to the unlabeled set (U). CNN-LapsELM performs better than ELM even without any labeled data.

Table 6.5 Performance comparison of various algorithms.

Evaluation index	DBN	CNN-SVM	AEM	CNN-ELM
RMSE	0.314	0.357	0.546	0.322
Accuracy	0.83	0.77	0.68	**0.82**
Training time	12.671	23.682	–	21.323
Testing time	6.893	7.723	–	8.217

Table 6.6 Performance comparison for various algorithms.

Evaluation index	CNN	ELM	SAE	SS-ELM	Ours
RMSE	0.423	0.402	0.368	0.353	0.309
Accuracy	0.69	0.74	0.76	0.79	0.87
Training time	21.582	8.579	13.257	22.315	25.237
Testing time	6.421	7.253	7.358	8.431	5.672

Accuracy, training time, and testing time were used to evaluate the performances of these algorithms. The RMSE index is used to evaluate the robustness of all the competitive algorithms. All the experiment results are described in Table 6.5. Results show that the proposed CNN-LapsELM has better performance than those of SVM and ELM in all training images, which also indicates that our proposed algorithm is able to fully leverage the unlabeled data to enhance the performances when compared to other supervised algorithms. Table 6.6 presents the testing times for CNN-LapsELM, SVM, and ELM. This shows that the testing performance is better than those of CNN, ELM, SAE, and SS-ELM because the Laplacian regularization can fully excavate the geometry information of sample distribution. However, the training time is relatively not improved, because the computation of Laplacian regularization term with all the available images is time-consuming.

6.5 Conclusions

In this framework, we proposed a novel algorithm CNN-LapsELM for industrial SD classification tasks. The main idea is fusing the deep features extracted from CNN and also apparent features extracted from frame hole images. Then we constructed semi-supervised ELM with Laplacian regularization, which was further used to classify the SD situation in industrial aluminum production cell. Compared to the previous studies, a brand-new Laplacian regularization semi-supervised ELM method was developed. The experimental result comparison demonstrated that the CNN-LapsELM algorithms outperform other existing methods. As the sensors and instruments have been abundantly applied to industries processes, our method can be further applied to various monitoring, fault diagnosis, and soft sensing domains.

References

[1] Z.X. Qiu, Principle and Application of Aluminum Electrolysis, China University of Mining and Technology Press, Xuzhou, 1998.
[2] Shui-ping Zeng, Jin-hong Li, Lei Ding, Fault diagnosis system for 350kA pre-baked aluminium reduction cell based on BP neural network, in: TMS–Light Metals, 2007.
[3] Marco A. Stam, Mark P. Taylor, John J.J. Chen, Albert Mulder, Renuka Rodrigo, Common behaviour and abnormalities in aluminium reduction cells, 2008, pp. 309–314.
[4] Jayson Tessier, Gary P. Tarcy, Eliezer Batista, Xiangwen Wang, Towards on-line monitoring of alumina properties at a pot level, in: Light Metals 2012, Springer, 2012, pp. 633–638.
[5] Weichao Yue, Xiaofang Chen, Weihua Gui, Yongfang Xie, Hongliang Zhang, A knowledge reasoning fuzzy-Bayesian network for root cause analysis of abnormal aluminum electrolysis cell condition, Frontiers of Chemical Science and Engineering 11 (3) (2017) 414–428.
[6] Weihua Gui, Weichao Yue, Xiaofang Chen, et al., Process industry knowledge automation and applications in aluminum reduction production process, Control Theory & Applications 35 (7) (2018) 887–899.
[7] Shuiping Zeng, Jinhong Li, Yuqian Wei, Danyang Cao, Calculation and control of equivalent superheat for 300kA prebake aluminum electrolysis, in: 2010 8th World Congress on Intelligent Control and Automation, IEEE, 2010, pp. 4755–4760.
[8] Zuguo Chen, Yonggang Li, Xiaofang Chen, Chunhua Yang, Weihua Gui, Semantic network based on intuitionistic fuzzy directed hyper-graphs and application to aluminum electrolysis cell condition identification, IEEE Access 5 (2017) 20145–20156.
[9] Yunsheng Liu, Shuyin Xia, Hong Yu, Yueguo Luo, Baiyun Chen, Kang Liu, Guoyin Wang, Prediction of aluminum electrolysis superheat based on improved relative density noise filter SMO, in: 2018 IEEE International Conference on Big Knowledge (ICBK), IEEE, 2018, pp. 376–381.
[10] Jingwen Zuo, Zhen Wang, Yang Zhang, Zhouquan Yan, Yali Zhao, Yuantao Chen, Research on image super-resolution algorithm based on mixed deep convolutional networks, Computers & Electrical Engineering 95 (2021) 107422.
[11] J.X. Zhao, Y.R. Qian, F.Z. Nan, et al., The method with CNN multilayer feature fusion and ELM diagnosis for breast diseases, Computer Engineering and Applications (2019) 1–7.
[12] Stefano Melacci, Mikhail Belkin, Laplacian support vector machines trained in the primal, Journal of Machine Learning Research 12 (3) (2011).
[13] Guang-Bin Huang, Qin-Yu Zhu, Chee-Kheong Siew, Extreme learning machine: a new learning scheme of feedforward neural networks, in: 2004 IEEE International Joint Conference on Neural Networks (IEEE Cat. No. 04CH37541), vol. 2, IEEE, 2004, pp. 985–990.
[14] Paul T. Sheeba, S. Murugan, Hybrid features-enabled dragon deep belief neural network for activity recognition, The Imaging Science Journal 66 (6) (2018) 355–371.

[15] Gengyan Zhao, Fang Liu, Jonathan A. Oler, Mary E. Meyerand, Ned H. Kalin, Rasmus M. Birn, Bayesian convolutional neural network based MRI brain extraction on nonhuman primates, NeuroImage 175 (2018) 32–44.
[16] Qing Shen, Xiaojuan Ban, Chong Guo, Cong Wang, Kernel semi-supervised extreme learning machine applied in urban traffic congestion evaluation, in: International Conference on Cooperative Design, Visualization and Engineering, Springer, 2016, pp. 90–97.
[17] Zhiqiong Wang, Mo Li, Huaxia Wang, Hanyu Jiang, Yudong Yao, Hao Zhang, Junchang Xin, Breast cancer detection using extreme learning machine based on feature fusion with CNN deep features, IEEE Access 7 (2019) 105146–105158.

CHAPTER 7

A novel semi-supervised soft sensor framework based on stacked auto-encoder wavelet extreme learning machine: SAE-WELM

7.1 Introduction

In modern processes industry, some key quality variables are hard to measure because they suffer the environmental limitations, expensive offline cost, and large delay. Thus many data-driven soft strategies have been proposed; see, for instance, [1–5]. The soft methods are mainly classified into three categories including black-box models, white-box models, and hybrid models (combining the black- and white-box models). The white-box models are based on the first principle, which limits the applications because it relies on knowledge of the process mechanism. With the development of a distributed control system, data-driven soft "black" models gradually gain more popularity because it is more flexible; for example, with some historical data, we can train a predictive model [6]. Researchers are likely to explore the nonlinear and dynamic function projection between the input variables and output key quality as an economic alternative to the complex and sophisticated process knowledge [7–10]. Most of the existing soft models use different nonlinear architecture to extract information contained in the processes data. Some popular models include principal component analysis (PCA) [11], deep belief network (DBN) [12], artificial neural network (ANN) [13], and probabilistic principal component regression (PPCR). Recently, some novel time-sequential models have been proposed. For example, Yuan et al. [5,14] proposed a soft sensor model based on supervised long-short time memories (LSTM), the application results showed that the supervised training performance outperforms the traditional fine-tuning approach. Besides, Xie et al. [13] developed a data-driven gated recurrent unit for soft sensor processing. Two stream factors are introduced in a gated recurrent

Learning-Based Soft Sensing and Predictions for Process Industries
https://doi.org/10.1016/B978-0-44-336759-5.00016-0

neural network, and the simulation study showed a good performance result. Ge et al. [11] created an efficient soft model with some process regression tasks. Chou [15] discussed the development of a physically soft sensor using sequence-to-sequence models. Jrvisalo [16] built a soft sensor model inflow rate prediction for specific energy estimation. Then Yuan et al. [10] introduced the hierarchical representation method for soft sensor domains. They employed an improved gradient descent to update the learning parameters. Other time sequence models of industrial processes have been discussed in [17,9].

Currently, with the deep architecture development, the performance of soft models based on deep learning have been greatly improved. Those models use the deep representation ability of input variables instead of shallow architecture, and the learning ability of complex tasks has been experimentally guaranteed. Further, with the sufficient processes, the hard environment and large amount of feature input in the processes industry, the deep learning techniques have shown their advantages. To obtain the complex nonlinear features, deep learning is considered as a remarkable frame to deal with these issues, and high accuracy and performance have been achieved [18–20,17,21–23,5,10,24]. The deep learning approaches have been extended to industrial process optimization, monitoring, fault diagnosis, biology, and chemistry domains. Compared to the shallow networks, the complex feature representation ability is limited, and the generalization ability is not appropriate to large processes systems [25,26,5]. The aforementioned results give us the motivation and inspiration to use the deep technique into the aluminum industry. However, these deep neural networks are mainly based on the background propagation solution; gradient diminishing and exploding issues are challenging in the training process. The computational complexity of these iterative methods have difficulties with generalization.

To improve the computational efficiency in the single-hidden-layer neural network, Huang et al. [27] proposed an extreme learning machine (ELM) with high performance in training the single-hidden layer network. ELM has been applied to many domains such as classification, regression tasks, pattern recognition, and industrial process domains. Many improved ELM-based methods have also been built. For example, Wong et al. [28] designed a multi-layer kernel-based ELM for representing features. Their model used different kernels and gained a generalized optimal model. Song et al. [29] proposed a semi-supervised ELM method, which combines the unlabeled data for the training process. Since then, some semi-supervised

ELM-based methods are also extended to soft sensor domains [6,11]. Lu [30] developed an online spatiotemporal ELM for complex time-varying distributed parameter systems, which can accurately model time-varying and nonlinear industrial processes. To take advantage of the deep feature extraction ability, the proposed method built the deep architectures for an aluminum electrolysis process. However, the traditional deep learning techniques use the back-propagation algorithms to train the whole network and have the shortcoming of the gradient diminishing and exploration, so in our network, we use the stacked autoencoders for the complex feature extraction and a top-layer extreme learning machine for the key quality identification. An ELM-based training method is also used for training the proposed soft model layer-by-layer. Some existing models use various activations to achieve better performances.

Deep kernel learning has recently attracted increasing attention in the industrial process modeling applications [31]. The main category of the deep kernel learning are: (1) the single-kernel model, which uses a single kernel as the nonlinear activation function, such as sigmoid, Gaussian, and tanh kernels; (2) the ensemble kernel model, which integrates two or more kernels to the deep network architectures. The radial basis function (RBF) is one of the most widely used models for learning the mapping function from the input to the output [32]. The common nonlinear activation function in RBF is the Gaussian kernel. However, the performance of RBF-based algorithms is largely decreased if the geometry distribution of input samples is not Gaussian. Further, the long tail in the Gaussian activation function from minus infinity to plus infinity limits the denoising ability. Some frequency of the input variables is also discarded. In contrast to the Gaussian kernel of RBF, a wavelet neural network is a highly efficient method for signal processing, and a shallow neural network has been proposed with the wavelet activation function. The shift factor enables the time axis of the wavelet neural network to perform ergodic analysis along with the temporal features, and scale parameter can shrink and stretch the wavelets to approximate the signals at different frequencies. Therefore, with different scale and shift transform, the better time–frequency localization property and focal characteristics are achieved. For example, Huang et al. [31,33] improved the traditional wavelet function and proposed a fuzzy wavelet neural network for prediction tasks. In [34] an improved wavelet neural network was built for electronic price prediction. This method considered the uncertainty factor, and a bootstrapping technique is introduced; however, the accuracy is easily influenced by different noises.

Zhang et al. [35] mathematically analyzed the nonlinear function, which has a better convergence rate and universal approximation ability. Papers [36,34,37,38] give the recent development, whereas their applications in the industrial plant process are not completely addressed, which partly motivates us for the current study.

To address the aforementioned problems, we proposed a hierarchical soft model to extract information and achieve a quality prediction for industrial processes. Wavelet activation was used to get the time-frequency location property of the input variables. First, the stacked auto-encoder was used for the complex feature extraction. Second, an ELM with a wavelet kernel activation function is built for a topmost classifier. Through this innovative combination, a more reliable and robust soft model can be generated. The different activation function is experimented for evaluating the performance of this algorithm. The main contributions of the paper are as follows:

The traditional deep architectures contain the issue of gradient diminishing and exploding. To improve the training efficiency and generalization ability, a novel deep soft model based on ELM with a wavelet kernel function is proposed.

A data-driven soft model was developed that can preserve the universal approximation ability and extract the deep and complex features in-process data. In contrast to the Gaussian kernel, the proposed approach combines the wavelet kernel in ELM, the generalization ability, and a greatly improved training speed.

To obtain a stable and applicable model, the approximation ability and convergence of the proposed SAE-WELM are proved by constructing a Lyapunov function.

7.2 Problem statement

The core problem can be stated as follows. Let x_l and x_u represent the training input samples with and without labels, respectively. The objective is to estimate the unknown target values y_u using the available data $X = x_l, x_u$ and the corresponding labels y_l associated with x_l. To achieve this, we aim to learn a nonlinear mapping function that predicts $\hat{y}_u$ based on the labeled and unlabeled data:

$$\hat{y}_u = f(x_l, x_u, y_l), \tag{7.1}$$

where $\hat{y}_u$ denotes the estimated values of the key quality-related target variable, and x represents the known input features.

7.3 Main framework

In this part, we propose a novel deep SAE-WELM method. We denote by $(\mathbf{X}, \mathbf{Y})$ the training samples. As shown in Fig. 7.1, a hierarchical auto-encoder and ELM for the top level are constructed. The current input of SAE-WELM is equal to the input data of X. The ith hidden layer output is further used as the $(i+1)$th input (layer-by-layer), and the whole network can be trained layer by layer. Differently from the traditional deep neural network, the continuous nonlinear wavelet activation function is applied. The deep SAE-WELM projects the input variables to a different dimension

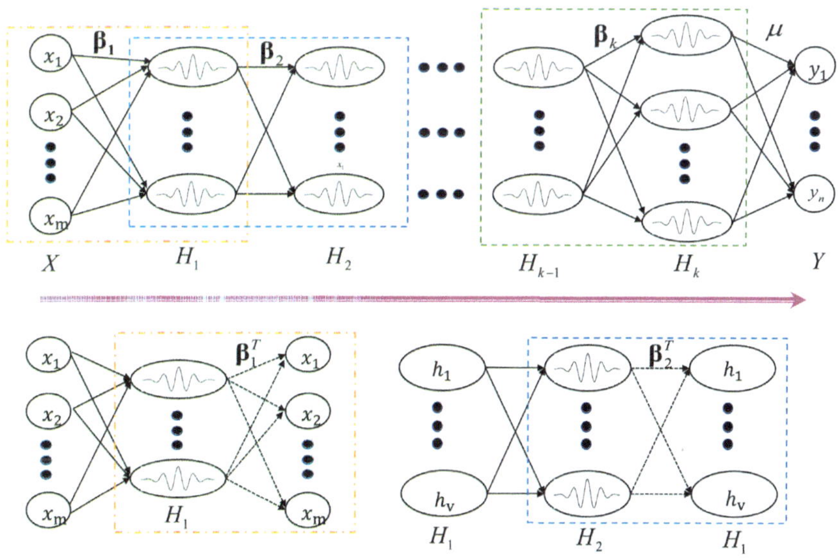

Figure 7.1 The flowchart of SAE-WELM.

space using the following equation:

$$h_i = \beta \cdot g(\varphi(x)) + b_i, \quad i = 1, \ldots, M, \tag{7.2}$$

where M is the number of hidden layers, g is the activation function, and φ is the wavelet kernel. The output matrix for the whole sample is presented by

$$H = \begin{pmatrix} g(\varphi_{a_1,b_1}(x_1)) & \cdots & g(\varphi_{a_L,b_L}(x_1)) \\ \vdots & \ddots & \vdots \\ g(\varphi_{a_1,b_1}(x_N)) & \cdots & g(\varphi_{a_L,b_L}(x_N)) \end{pmatrix}, \tag{7.3}$$

where a is the scaling translation, and b is the dilation parameter, x is the number of training samples, N is the number of input nodes, and L is the number of hidden neurons.

The projection output matrix is represented as follows:

$$H_j = g\left((\beta_j)^T H_{j-1} \right), \qquad i = 1, \ldots, K. \tag{7.4}$$

The output weights β are computed by the Moore–Penrose pseudoinverse, which meets the following constraints:

$$\begin{gathered} \beta = \mathbf{H}^{-1}\mathbf{X} \\ s.t.\ \beta\beta^T = \mathbf{I}, \end{gathered} \tag{7.5}$$

where I is an identity matrix whose square elements are 1s, and other elements are 0s. The implementation of the SAE-WELM can be given in the following steps.

(1) Choose the related input variables according to the experienced worker and operators. All collected data are pre-processed by normalization and standardization.

(2) Divide the dataset patterns with training, validation, and testing samples.

(3) Determine the architecture of the proposed SAE-WELM, the number of the hidden layer, and the neurons in each layer. Initialization of the shift a, dilation parameters b, and residual error threshold ε.

(4) Train the hierarchical autoencoder layer-by-layer with Eq. (7.4) and gain the output of K hidden layers of $H(1) - H(K)$.

(5) Train the highest ELM with Eq. (7.5) for obtaining the output weights of the SAE-WELM.

(6) Evaluate the performance of the proposed soft models with the testing patterns. If the residual error is lower than the threshold, then we give the projection of the SAE-WELM. Otherwise, update the dilation parameter with the candidates and then go back to Step (4).

(7) Apply the trained SAE-WELM for key quality identification to obtain an output of the layer weights and the nonlinear projection function.

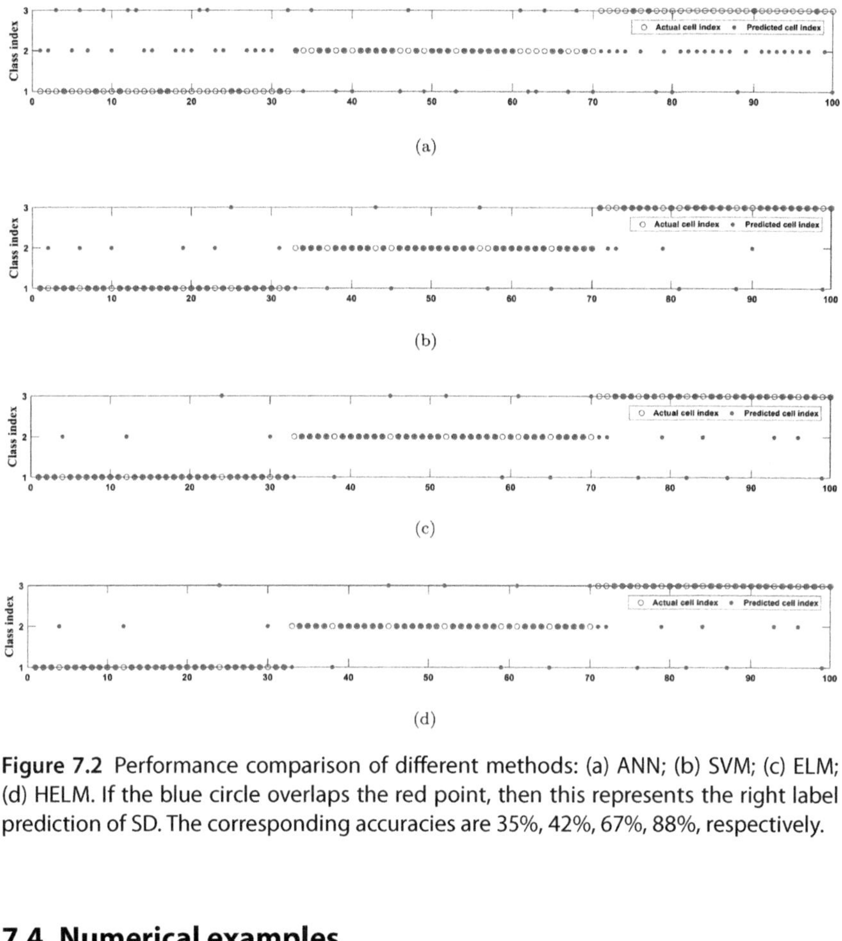

Figure 7.2 Performance comparison of different methods: (a) ANN; (b) SVM; (c) ELM; (d) HELM. If the blue circle overlaps the red point, then this represents the right label prediction of SD. The corresponding accuracies are 35%, 42%, 67%, 88%, respectively.

7.4 Numerical examples

The validation process is based on the collected samples and is divided into training, validation, and testing sets. From the 100 testing patterns performance comparison, the SAE-WELM has a significant accuracy improvement than the traditional ANN. The results demonstrate that some deep architecture (HELM, DBN) has a better representation learning ability than the shallow network (ANN). Compared with Fig. 7.2, the average testing accuracy and training time of different algorithms with different hidden units are given. The results show that the accuracy is relatively stable with the advantage of shorter training time, whereas RBF with Gaussian kernel is not more robust than SAE-WELM. A low RMSE value indicates a superior performance (see Fig. 7.3). We can see that the accuracies of these methods are generally improved with the increment of the hidden units in both ANN and RBF. However, the training time also tends to increase, which means that more computation cost is necessary for building a high-dimension neural network. Due to the fast metric closed solution of the proposed SAE-WELM, the training speed is clearly better than BP-based ANN and RBF. A good performance can be shown in these figures. The ANN is the most time-consuming method, which is the result of BP-based iterative steps. Our method is competitive in time complexity with RBF network; however, the accuracy outperforms the RBF-based networks.

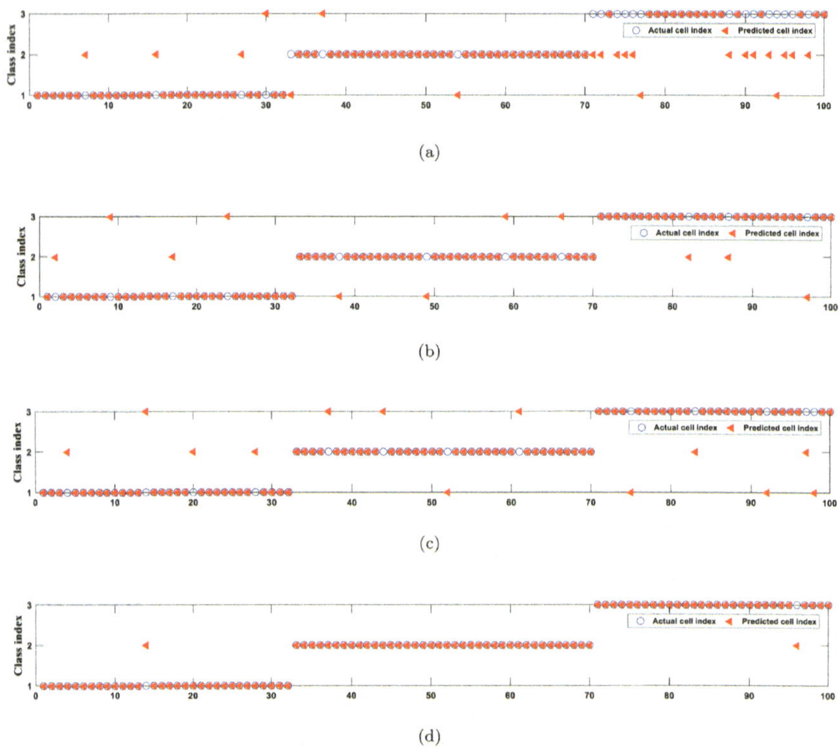

Figure 7.3 Performance comparison of various kernels of algorithms: (a) Gaussian RBF; (b) DBN; (c) Sigmoid (SAE-ELM); (d) Ours (SAE-WELM). The corresponding accuracies are 79%, 89%, 87%, 98%, respectively.

The structure parameters are given in Table 7.1. To make a more detailed robustness analysis of the proposed method, Table 7.2 presents the results with various hidden layer units.

Compared with Fig. 7.4, the average testing accuracy and training time of various algorithms with various hidden units are given. The results show

Table 7.1 The structure parameters of various algorithms.

Methods	(K)	ith (V_i)	Hidden units	Initialization
ELM	–	–	200	$C = 25$
ANN	–	–	200	learning rate 0.02
RBF	–	–	200	learning rate 0.02
DBN	3	10	200	–
SAE-WELM	3	10	200	$a = 4.5$, $b = 0$

Table 7.2 Training times and variances of various algorithms with various numbers of hidden nodes. A: Variance, B: Training time (s), C: Testing RMSE.

N_h	ELM			RBF			SAE-WELM			DBN		
	A	B	C	A	B	C	A	B	C	A	B	C
100	0.0231	2.3	0.0891	0.1235	2.4	0.0812	0.0083	**0.038**	0.0076	0.0842	0.25	0.0282
200	0.0121	2.8	0.0742	0.0832	4.2	0.0563	0.0083	**0.034**	**0.0023**	0.0768	0.59	0.0187
400	0.0109	3.5	0.0653	0.0756	5.5	0.0481	0.0074	**0.042**	0.0215	0.0746	0.71	0.0153
600	0.0098	4.9	0.0649	0.0678	6.3	0.0257	**0.0071**	**0.047**	0.0157	0.0654	0.93	0.0125

that the accuracy is relatively stable with the advantage of a less training time, whereas RBF with Gaussian kernel is not more robust than SAE-WELM. We can see that the accuracies of these methods are generally improved with the increment of the hidden units in both ANN and RBF. However, the training time is also an increasing tendency, which means that more computation cost is necessary for building a high-dimension neural network. Due to the fast matrix closed solution of the proposed SAE-WELM, the training speed is clearly better than those of BP-based ANN and RBF. A good performance can be seen in these figures. The ANN is the most time-consuming method, which is the result of BP-based iterative steps. Our method is competitive in time complexity with RBF network; however, the accuracy outperforms the RBF-based networks.

To make a comparison, RBF, ANN, and SAE-WELM are set as a competitive rival. The intuition shows that the proposed method is not sensitive to manual selection of user-specified parameters. From these experimental results, the proposed soft model has a good performance robustness in SD identification of the aluminum electrolysis plant. Table 7.2 gives the related index of variance to evaluate the robustness; compared to traditional ELM and RBF methods, the proposed SAE-WELM has the competitive advantage in robustness. Further, with the increase of the hidden nodes, the variance decreased, which means that the proposed SAE-WELM has a better generalization performance. Based on the former experiments in the aluminum electrolysis plant application, the To To sum up, SAE-WELM has achieved the following significant performances.

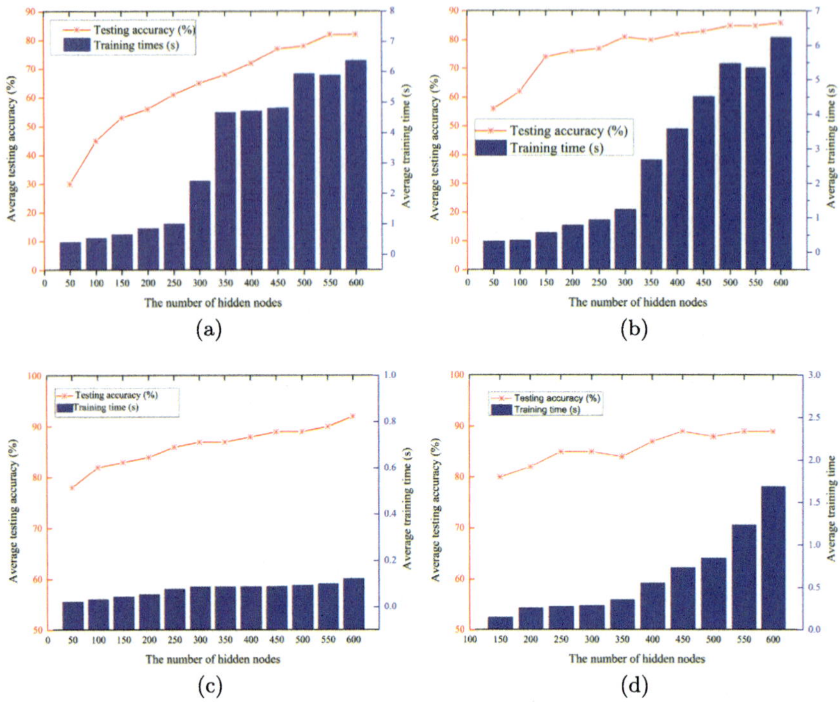

Figure 7.4 Performance comparisons with various hidden layer nodes: (a) ANN, (b) RBF, (c) SAE-WELM, (d) DBN.

Excellence performance in the wide-plant SD identification: Since the SD classification in industrial aluminum production is a challenging work for operators and technicians, an efficient and intelligent deep scheme is necessary to replace the manual detection, which has the difficulty of subjectivity. The industrial experiments showed that the proposed SAE-WELM provided a superior SD accuracy.

Reduction of the computation cost: The traditional methods are back-propagation-based neural methods, which have the issue of gradient diminishing and explosion. The proposed SAE-WELM learns the hidden weights by a closed matrix solution, so the computational efficiency is hundreds of times faster than the iterative approaches. The proposed method can be directly used for the plant SD classification as an alternative to the human and hence saves cost and increases the profit.

7.5 Conclusions

In this framework, we developed an effective and efficient soft model based on deep extreme learning machine and wavelet activation function. This model reserves universal approximation ability and fast training speed. Industrial aluminum plant-wide production demonstrated that the proposed SAE-WELM has good robustness and generalization ability. In principle, the proposed soft model can be extended to other plant-wide industrial process applications. Furthermore, the following topics deserve a further discussion: (1) Wavelet is a stable bandwidth filter, and some multi-resolution analysis (MRA) of the samples deserve to be discussed. The time frequency information of the related input variables can be used for training the SAE-WELM; (2) The flexibility of the proposed method can be further discussed, and an adaptive neural network can be built to optimize the proposed SAE-WELM architecture; (3) The SAE-ELM performance relies on the manual feature selection. To deal with the noisy data, the performance of the proposed wavelet-based soft model can be further explored.

References

[1] Qiumei Cong, Wen Yu, Integrated soft sensor with wavelet neural network and adaptive weighted fusion for water quality estimation in wastewater treatment process, Measurement 124 (2018) 436–446.

[2] Divish Rengasamy, Mina Jafari, Benjamin Rothwell, Xin Chen, Grazziela P. Figueredo, Deep learning with dynamically weighted loss function for sensor-based prognostics and health management, Sensors 20 (3) (2020) 723.

[3] Ruimin Xie, Kuangrong Hao, Biao Huang, Lei Chen, Xin Cai, Data-driven modeling based on two-stream λ gated recurrent unit network with soft sensor application, IEEE Transactions on Industrial Electronics 67 (8) (2019) 7034–7043.

[4] Weiwu Yan, Di Tang, Yujun Lin, A data-driven soft sensor modeling method based on deep learning and its application, IEEE Transactions on Industrial Electronics 64 (5) (2016) 4237–4245.

[5] Xiaofeng Yuan, Lin Li, Yuri A.W. Shardt, Yalin Wang, Chunhua Yang, Deep learning with spatiotemporal attention-based lstm for industrial soft sensor model development, IEEE Transactions on Industrial Electronics 68 (5) (2020) 4404–4414.

[6] Le Yao, Zhiqiang Ge, Deep learning of semisupervised process data with hierarchical extreme learning machine and soft sensor application, IEEE Transactions on Industrial Electronics 65 (2) (2017) 1490–1498.

[7] Bingbing Shen, Le Yao, Zhiqiang Ge, Nonlinear probabilistic latent variable regression models for soft sensor application: from shallow to deep structure, Control Engineering Practice 94 (2020) 104198.

[8] Xiaofeng Yuan, Chen Ou, Yalin Wang, Chunhua Yang, Weihua Gui, A layer-wise data augmentation strategy for deep learning networks and its soft sensor application in an industrial hydrocracking process, IEEE Transactions on Neural Networks and Learning Systems 32 (8) (2019) 3296–3305.

[9] Xiaofeng Yuan, Shuaibin Qi, Yalin Wang, Haibing Xia, A dynamic CNN for nonlinear dynamic feature learning in soft sensor modeling of industrial process data, Control Engineering Practice 104 (2020) 104614.

[10] Xiaofeng Yuan, Jiao Zhou, Biao Huang, Yalin Wang, Chunhua Yang, Weihua Gui, Hierarchical quality-relevant feature representation for soft sensor modeling: a novel deep learning strategy, IEEE Transactions on Industrial Informatics 16 (6) (2019) 3721–3730.

[11] Jinlin Zhu, Zhiqiang Ge, Zhihuan Song, Distributed parallel pca for modeling and monitoring of large-scale plant-wide processes with big data, IEEE Transactions on Industrial Informatics 13 (4) (2017) 1877–1885.

[12] Yalin Wang, Zhuofu Pan, Xiaofeng Yuan, Chunhua Yang, Weihua Gui, A novel deep learning based fault diagnosis approach for chemical process with extended deep belief network, ISA Transactions 96 (2020) 457–467.

[13] Shiwen Xie, Yongfang Xie, Hao Ying, Zhaohui Jiang, Weihua Gui, Neurofuzzy-based plant-wide hierarchical coordinating optimization and control: an application to zinc hydrometallurgy plant, IEEE Transactions on Industrial Electronics 67 (3) (2019) 2207–2219.

[14] Xiaofeng Yuan, Lin Li, Yalin Wang, Nonlinear dynamic soft sensor modeling with supervised long short-term memory network, IEEE Transactions on Industrial Informatics 16 (5) (2019) 3168–3176.

[15] Cheng-Hung Chou, Haibin Wu, Jia-Lin Kang, David Shan-Hill Wong, Yuan Yao, Yao-Chen Chuang, Shi-Shang Jang, John Di-Yi Ou, Physically consistent soft-sensor development using sequence-to-sequence neural networks, IEEE Transactions on Industrial Informatics 16 (4) (2019) 2829–2838.

[16] Markus Järvisalo, Tero Ahonen, Jero Ahola, Antti Kosonen, Markku Niemelä, Soft-sensor-based flow rate and specific energy estimation of industrial variable-speed-driven twin rotary screw compressor, IEEE Transactions on Industrial Electronics 63 (5) (2016) 3282–3289.

[17] Qingqiang Sun, Zhiqiang Ge, Probabilistic sequential network for deep learning of complex process data and soft sensor application, IEEE Transactions on Industrial Informatics 15 (5) (2018) 2700–2709.

[18] Feng Duan, Lili Dai, Wennan Chang, Zengqiang Chen, Chi Zhu, Wei Li, sEMG-based identification of hand motion commands using wavelet neural network combined with discrete wavelet transform, IEEE Transactions on Industrial Electronics 63 (3) (2015) 1923–1934.

[19] S. Kolås, T. Støre, Bath temperature and AlF3 control of an aluminium electrolysis cell, Control Engineering Practice 17 (9) (2009) 1035–1043.

[20] Mehdi Rafiei, Taher Niknam, Mohammad-Hassan Khooban, Probabilistic forecasting of hourly electricity price by generalization of ELM for usage in improved wavelet neural network, IEEE Transactions on Industrial Informatics 13 (1) (2016) 71–79.

[21] Yong-fang Xie, Shi-wen Xie, Yong-gang Li, Chun-hua Yang, Wei-hua Gui, Dynamic modeling and optimal control of goethite process based on the rate-controlling step, Control Engineering Practice 58 (2017) 54–65.

[22] Hao-Fan Yang, Tharam S. Dillon, Elizabeth Chang, Yi-Ping Phoebe Chen, Optimized configuration of exponential smoothing and extreme learning machine for traffic flow forecasting, IEEE Transactions on Industrial Informatics 15 (1) (2018) 23–34.

[23] Xianhui Yin, Zhanwen Niu, Zhen He, Zhaojun Steven Li, Dong-hee Lee, Ensemble deep learning based semi-supervised soft sensor modeling method and its application on quality prediction for coal preparation process, Advanced Engineering Informatics 46 (2020) 101136.

[24] Yi Zheng, Ning Li, Shaoyuan Li, Hot-rolled strip laminar cooling process plant-wide temperature monitoring and control, Control Engineering Practice 21 (1) (2013) 23–30.
[25] Yongxiang Lei, Xiaofang Chen, Ming Min, et al., A semi-supervised Laplacian extreme learning machine and feature fusion with CNN for industrial superheat identification, Neurocomputing 381 (2020) 186–195.
[26] Yongxiang Lei, Xiaofang Chen, Yongfang Xie, An improved cell situation identification approach with convolutional neural network and wavelet extreme learning machine, in: Proceedings of the Institution of Mechanical Engineers, Part I: Journal of Systems and Control Engineering, 2020, 0959651820935667.
[27] Jiexiong Tang, Chenwei Deng, Guang-Bin Huang, Extreme learning machine for multilayer perceptron, IEEE Transactions on Neural Networks and Learning Systems 27 (4) (2015) 809–821.
[28] Chi Man Wong, Chi Man Vong, Pak Kin Wong, Jiuwen Cao, Kernel-based multilayer extreme learning machines for representation learning, IEEE Transactions on Neural Networks and Learning Systems 29 (3) (2016) 757–762.
[29] Gao Huang, Shiji Song, Jatinder N.D. Gupta, Cheng Wu, Semi-supervised and unsupervised extreme learning machines, IEEE Transactions on Cybernetics 44 (12) (2014) 2405–2417.
[30] Xinjiang Lu, Feng Yin, Chang Liu, Minghui Huang, Online spatiotemporal extreme learning machine for complex time-varying distributed parameter systems, IEEE Transactions on Industrial Informatics 13 (4) (2017) 1753–1762.
[31] Wei Huang, Sung-Kwun Oh, Witold Pedrycz, Fuzzy wavelet polynomial neural networks: analysis and design, IEEE Transactions on Fuzzy Systems 25 (5) (2016) 1329–1341.
[32] Jooyoung Park, Irwin W. Sandberg, Universal approximation using radial-basis-function networks, Neural Computation 3 (2) (1991) 246–257.
[33] Wei Huang, Sung-Kwun Oh, Witold Pedrycz, Hybrid fuzzy wavelet neural networks architecture based on polynomial neural networks and fuzzy set/relation inference-based wavelet neurons, IEEE Transactions on Neural Networks and Learning Systems 29 (8) (2017) 3452–3462.
[34] Mehdi Rafiei, Taher Niknam, Jamshid Aghaei, Miadreza Shafie-Khah, João P.S. Catalão, Probabilistic load forecasting using an improved wavelet neural network trained by generalized extreme learning machine, IEEE Transactions on Smart Grid 9 (6) (2018) 6961–6971.
[35] Jun Zhang, Gilbert G. Walter, Yubo Miao, Wan Ngai Wayne Lee, Wavelet neural networks for function learning, IEEE Transactions on Signal Processing 43 (6) (1995) 1485–1497.
[36] Soheil Ganjefar, Morteza Tofighi, Hamidreza Karami, Fuzzy wavelet plus a quantum neural network as a design base for power system stability enhancement, Neural Networks 71 (2015) 172–181.
[37] Shiwen Xie, Yongfang Xie, Weihua Gui, Chunhua Yang, Weighted-coupling CSTR modeling and model predictive control with parameter adaptive correction for the goethite process, Journal of Process Control 68 (2018) 254–267.
[38] Sevcan Yilmaz, Yusuf Oysal, Fuzzy wavelet neural network models for prediction and identification of dynamical systems, IEEE Transactions on Neural Networks 21 (10) (2010) 1599–1609.

CHAPTER 8

A novel soft senor based on Laplacian–Hessian semi-supervised hierarchical extreme learning machine: LHSS-HELM

8.1 Introduction

In modern industrial processes, some key variables represent important indicators related to product quality and production efficiency. However, it is quite difficult to measure most of these variables using physical instruments because of the strict and harsh limitations of measurement environments, high offline analyzer cost, and large delay. A soft sensor is a virtual sensing technique to estimate hard-to-measure key variables through certain easy-to-measure process variables [1–3]. It provides a feasible and economical alternative to costly and impractical physical measurement devices. In addition, with the rapid development of distributed control systems, many data-driven soft sensor modeling techniques have been extensively studied and applied successfully to many domains with time savings and low cost, particularly in process monitoring [4,5]. Many data-driven soft modeling methods, such as artificial neural networks (ANNs) [6], principal component analysis (PCA) [7], and partial least squares (PLS) [8], have extended the popularity of soft sensors. However, they can only use labeled data that contain both input and output samples, and extensive unlabeled data are discarded.

In the last decade, Huang et al. proposed the extreme learning machine (ELM) as a new training algorithm for the single hidden feedforward network (SLFN) [9]. Compared with conventional approaches, ELM only solves an analytic matrix about the output weights between the input and hidden layers, whereas the input weights and hidden neurons are randomly generated. By using the squared loss of the prediction, the output weights training process can be translated into a regularized least (or ridge) problem, the solution for which is a closed form. Many recent studies

Learning-Based Soft Sensing and Predictions for Process Industries
https://doi.org/10.1016/B978-0-44-336759-5.00017-2

have indicated that ELM offers comparable or even better performance in many classification and regression tasks [10–14]. Lei et al. [14] investigated the distributed convex optimization under persistent attacks with a hybrid system. ELM can also be extended to many specific applications with high computation efficiency and generalization ability [15–19]. Some other efficient artificial models, such as fuzzy logic [7], genetic algorithms [8], and convolutional neural network (CNN) [9], have been widely used in the application domains.

In recent years, deep learning (DL) has become a hot topic for complex feature extraction in process modeling. A layer-by-layer unsupervised learning method with deep networks has been used in many soft modeling processes. DL methods extend single hidden layer networks and use multilayer architecture to achieve remarkable feature representation. It has been confirmed that DL architectures such as deep soft networks [19–22] and long–short time memory (LSTM) [23], which are especially suited for processing soft sensors, have exhibited excellent performance. However, these DL methods are mostly trained by a backpropagation (BP) algorithm, and all the parameters must be fine-tuned for the entire deep network, which is commonly a time-consuming process. Wong [24] considered the outlier in constructing a roller bearing degradation assessment method based on a deep MLP convolution neural network. Min et al. [16] proposed a semi-supervised hierarchical ELM (SS-HELM) that significantly decreases computational cost as compared to the traditional DL methods in terms of quality variable prediction tasks. Recently, DL modeling in industrial processes was introduced for soft sensor and monitoring tasks [25–28]. In addition, Laplacian regularization was incorporated as a regularization term to penalize the objective function [29]. However, this Laplacian-based method lacks extrapolating ability and may exhibit fast fading performance when only a few labeled data are available. Yuan et al. proposed a DL feature representation based on variable-wise weighted stacked autoencoder (SAE), which enhances the prediction performance as compared to traditional multilayer neural network (NN) and SAE. Lei et al. analyzed the performance difference of shallow neural network toward deep architectures and also built the latent variables model for regression prediction. Lei et al. [30] investigated the inner knowledge of the aluminum production and built a semi-supervised extreme learning method. They fused the deep complex features and density features in training image samples. Although several literature sources have been published, DL is in its initial stage for soft sensor domains, some problems are still unsolved, and more studies are thus required. For example, based on Laplacian regularization, SS-ELM can determine the geometric structure of underlying manifolds even as it lacks extrapolating power and biases of the solution toward a constant.

In this study, a deep hybrid regularization semi-supervised architecture is proposed for industrial soft sensor modeling processes. We can use all labeled and unlabeled data to establish the entire modeling. First, the deep autoencoder (AE) structure is used to extract features included in all available process data using an unsupervised hierarchical learning method. Then a semi-supervised ELM is further used in quality variable prediction with all available process data, including both labeled and unlabeled data. In addition, Laplacian and Hessian regularization methods are integrated into semi-supervised soft modeling training to improve the extrapolating power and robustness. Fu et al. introduce a single-layer hybrid network for superheat degree classification in aluminum electrolysis production. In this chapter, we propose a deep extended efficient autoencoders network for the cell temperature (CT) regression. In particular, the following contributions are made.

- An efficient soft sensor modeling algorithm that can significantly improve computational efficiency is proposed. The developed model constructs a deep architecture that fully enhances the Laplacian and Hessian terms for improving the extrapolating ability and robustness.
- By adding unlabeled data the accuracy of the proposed semi-supervised soft sensor model outperforms traditional supervised soft sensor models.
- Different from the study by Shen et al., a deep AE is built for extracting complex features and learning in its hidden weights by the layer-to-layer low-computation training alternatives.

8.2 Problem statement

The CT is a vital indicator that can reflect the whole production situation of the aluminum electrolysis process. Keeping the CT in a suitable range can guarantee the highly efficient production and extend the life span of the bath. However, in the practical production, the detection of CT is mainly based on the thermocouple, which is a device with expensive cost and time delay. Furthermore, the method of detection measures the local surface temperature, which cannot reflect the whole field of the aluminum production. An analysis of the knowledge mechanism indicates that the inner links exist in these processes variables, which means that some data-driven model can be proposed.

Denote by $\mathbf{x}$ the input variables and by $\mathbf{y}$ and $\hat{\mathbf{y}}$ the labeled and predicted targets. In semi-supervised learning, two fundamental assumptions are generally made:

1. The labeled data X_l and unlabeled data X_u are drawn from the same marginal distribution P_X.
2. If two samples x_i and x_j are close in the feature space, then the corresponding conditional distributions $P(y|x_i)$ and $P(y|x_j)$ are also similar.

The second assumption, often referred to as the *smoothness assumption*, enables us to enforce regularity over the data manifold. Based on this, the Laplacian regularization term can be introduced as

$$L_m = \frac{1}{2} \sum_{i,j} u_{ij} \left\| P(y|x_i) - P(y|x_j) \right\|^2, \tag{8.1}$$

where u_{ij} denotes the similarity between samples x_i and x_j, typically calculated using the Gaussian kernel

$$u_{ij} = \exp\left(-\frac{\|x_i - x_j\|^2}{2\sigma^2} \right) \tag{8.2}$$

or set to 1 if x_i and x_j are neighbors.

The regularization term L_m can be written in matrix form using the graph Laplacian

$$\hat{L}_m = \mathrm{Tr}(\hat{Y}^\top L \hat{Y}), \tag{8.3}$$

where $\hat{Y}$ is the predicted output matrix, $\mathrm{Tr}(\cdot)$ denotes the trace operator, and $L = D - U$ is the graph Laplacian matrix with similarity matrix $U = [u_{ij}]$ and diagonal matrix D with entries $D_{ii} = \sum_j u_{ij}$.

This Laplacian term promotes similar predictions for neighboring data points in the feature space, encouraging smoothness over the manifold. However, it is known that Laplacian regularization may bias solutions toward constants and lacks strong extrapolating ability in cases with few labeled data. This motivates the integration with Hessian regularization to enhance robustness and generalization.

8.3 Main framework

The structure of the proposed LHSS-HELM is presented in Fig. 8.1. First, we can train the HAEs to obtain multilayer weights $\mathbf{W}_i\ (i = 1, \ldots, h)$ in an unsupervised learning method with all the available data. The parameters of the AE are obtained as an ELM method as given in Eq. (8.4). All the hidden weights parameters are obtained layer-by-layer. To extract deep and complex features, HAEs are constructed, and the parameters are trained layer-by-layer. Compared to the BP algorithm, the network parameters are obtained in a closed solution rather than in an iterative way, and then the computational complexity is greatly improved. Furthermore, for top-layer HAE-based ELM training, Laplacian and Hessian regularizations are integrated to take advantage of the manifold structure of the data space to improve the performance of the traditional ELM.

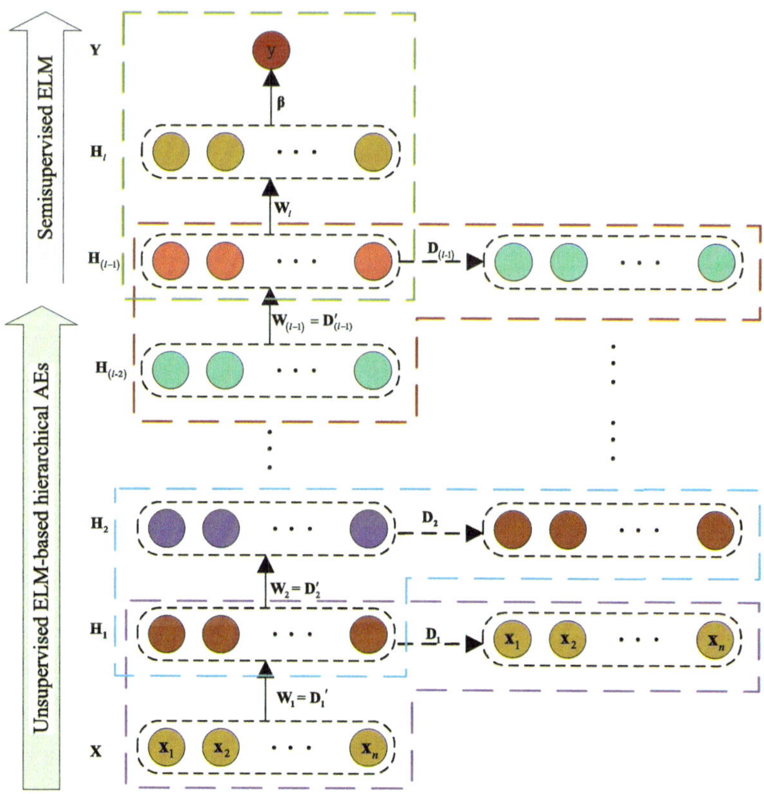

Figure 8.1 LHSS-HELM network structure.

The loss function of the proposed framework is then formulated as follows:

$$
\begin{aligned}
\Gamma_{\text{LHSS-ELM}} = & \min_{\boldsymbol{\beta} \in R^{n_h \times n_o}} \frac{1}{2}\|\boldsymbol{\beta}\|^2 + \frac{C}{2}\|\mathbf{e}_i\|^2 \\
& + \frac{\lambda_1}{2}\operatorname{Tr}\left(\hat{\mathbf{Y}}^T\mathbf{L}\hat{\mathbf{Y}}\right) + \frac{\lambda_2}{2}\operatorname{Tr}\left(\hat{\mathbf{Y}}^T\mathbf{H}\hat{\mathbf{Y}}\right) \\
\text{s.t.} \quad & \mathbf{e}_i = \mathbf{y}_i^T - \mathbf{J}(\mathbf{x}_i)\boldsymbol{\beta}, i = 1, \ldots, l, \\
& \hat{\mathbf{Y}}_i = \mathbf{J}(\mathbf{x}_i)\boldsymbol{\beta}, i = 1, \ldots, l+u,
\end{aligned} \tag{8.4}
$$

where $\mathbf{L} \in \mathbb{R}^{(l+u)\times(l+u)}$ and $\mathbf{H} \in \mathbb{R}^{(l+u)\times(l+u)}$ are Laplacian and Hessian matrices derived from labeled and unlabeled data, respectively, $\mathbf{Y}$ is the output matrix, $\boldsymbol{\beta} \in \mathbb{R}^{n_h \times n_o}$ represents the output weights, and λ_1 and λ_2 are trade-off parameters. In our framework, we associate different coefficients C_i on prediction errors with examples from skewed classes. As studied in [31], we alleviate poor generalization by using the penalty coefficient of $C_i{=}C_0/N_{t_i}$, where C_0 is a user-defined parameter as in conventional ELM, and N_t is the number of training patterns in class t_i. Therefore the samples from the dominant classes will not be overfitted, whereas the patterns from the less dominant class examples will not be neglected simultaneously. By substituting the constraints into the objective function we rewrite (8.4) in the following matrix form:

$$
\begin{aligned}
\Gamma_{\text{LHSS-ELM}} = & \min_{\boldsymbol{\beta} \in R^{n_h \times n_o}} \frac{1}{2}\|\boldsymbol{\beta}\|^2 + \frac{C}{2}\left\|\mathbf{Y} - \mathbf{J}\boldsymbol{\beta}\right\|^2 \\
& + \frac{\lambda_1}{2}\operatorname{Tr}\left(\boldsymbol{\beta}^T\mathbf{J}^T\mathbf{L}\mathbf{J}\boldsymbol{\beta}\right) \\
& + \frac{\lambda_2}{2}\operatorname{Tr}\left(\boldsymbol{\beta}^T\mathbf{J}^T\mathbf{H}\mathbf{J}\boldsymbol{\beta}\right),
\end{aligned} \tag{8.5}
$$

where $\hat{\mathbf{Y}} \in \Re^{(l+u)\times n_o}$ is an augmented training target whose first l rows are equal to $\mathbf{Y}_l$, and the remaining rows are set to 0, and $\mathbf{C}$ is the $(l+u)\times(l+u)$ diagonal matrix with first l diagonal elements $[\mathbf{C}]_{ii} = C_i$ and the remaining elements equal to 0.

By setting the gradient of (8.5) to 0 we can calculate the output weights of LHSS-ELM:

$$
\boldsymbol{\beta}^* = \left(\mathbf{I}_{n_h} + \mathbf{J}^{\mathrm{T}}\mathbf{C}\mathbf{J} + \lambda_1\mathbf{J}^{\mathrm{T}}\mathbf{L}\mathbf{J} + \lambda_2\mathbf{J}^{\mathrm{T}}\mathbf{H}\mathbf{J}\right)^{-1}\mathbf{J}^T\mathbf{C}\mathbf{Y}, \tag{8.6}
$$

where $\mathbf{I}_{n_h}$ is the $n_h \times n_h$ identity matrix. When the number of labeled data is less than the number of hidden neurons, the alternative solution of LHSS-

ELM is

$$\beta^* = \mathbf{J}^T(\mathbf{I}_{l+u} + \mathbf{C}\mathbf{J}\mathbf{J}^T + \lambda_1 \mathbf{L}\mathbf{J}\mathbf{J}^T + \lambda_2 \mathbf{H}\mathbf{J}\mathbf{J}^T)^{-1}\mathbf{C}\mathbf{Y}, \tag{8.7}$$

where $\mathbf{I}_{(l+u)}$ is the $(l+u) \times (l+u)$ identity matrix. The detailed procedure is given in **Algorithm 1.**

Algorithm 1 LHSS-HELM algorithm.

Require: Labeled data $\{\mathbf{X}_l, \mathbf{Y}_l\} = \{\mathbf{x}_i, \mathbf{y}_i\}_{i=1}^{l}$;
Unlabeled data $\mathbf{X}_u = \{\mathbf{x_i}\}_{i=1}^{u}$;
Penalty coefficient C_0 and trade-off parameter sequence λ;

Ensure: The modeling mapping function $\widehat{\mathbf{Y}} = f(\mathbf{x})$; The output prediction of testing patterns $\widehat{\mathbf{Y}}$.

Start:

step 1: Select the variables from the technician workers' experience

step 2: Preprocess the data (normalization, standardization);

step 3: Split data into training, validation, and testing data;

step 4: Determine the number of HAEs and of neurons in each layer;

step 5: Randomly initialize input weights, train HAEs with all the available input data, and obtain weight $\mathbf{W}_i$ for each layer;

step 6: Construct the Laplacian term L and Hessian operator H from both labeled and unlabeled $\mathbf{X}$;

step 7: Initiate an ELM framework of n_h hidden neurons with random weights;

step 8: Calculate the output of the hidden neurons of ELM $\mathbf{J} \in \Re^{(l+u)\times n_h}$;

step 9: **If** $n_h \le N$, **then** compute the output weights $\boldsymbol{\beta}$ using (8.6);
Else compute the output weights $\boldsymbol{\beta}$ using (8.7);

step 10: Test whether the recent model is the best model:
 (a) if **yes**, then compute the modeling mapping function $\widehat{\mathbf{Y}} = f(\mathbf{x}, \mathbf{W}_i\,(i = 1, \ldots, h), \boldsymbol{\beta})$;
 (b) if **no**, then update λ and return to step 8;

step 11: Predict the quality variable $\widehat{\mathbf{Y}}$ for testing patterns.

end.

8.4 Numerical examples

A performance comparison using various unlabeled samples was conducted on the LHSS-HELM model to demonstrate that the extra scale unlabeled samples would greatly affect the prediction performance and robustness.

We gradually increased the unlabeled data from 0 to 1000 to train the ELM, SS-ELM, HELM, and LHSS-HELM models. 400 samples from the testing dataset were used to evaluate the prediction results. ELM and HELM employed 500 labeled samples for the supervised learning, and SS-ELM and LHSS-HELM are conducted under the same conditions for semi-supervised learning. This means that both the performance and robustness are improved with the additional unlabeled training data. Some similar results could also be observed in the comparison of the basic ELM, HELM, and SS-ELM methods when only labeled training data are used, whereas LHSS-ELM degenerated to a universal ELM. This indicates that the proposed algorithm with additional unlabeled samples can effectively improve the classification performance and robustness. In addition, the training time was slightly longer because of the use of Hessian operators, see Tables 8.1 and 8.2.

Table 8.1 Parameters set for ELM, SS-ELM, HELM, and LHSS-HELM.

Methods	L_{sae}	C	λ_1	λ_2	N_{elm}	N_{sae}
ELM	–	5	0.1	0.2	100	–
SS-ELM	–	5	0.1	0.2	100	–
SS-HELM	3	25	0.1	0.2	200	10
LHSS-HELM	3	25	0.1	0.2	200	10

Table 8.2 Comparison results with various methods. A: Testing RMSE, B: Training time (s), C: Variance.

Methods	**A**	**B**	**C**
SS-PPCR	0.0069	16.91	0.0078
Co-PLS	0.0062	15.24	0.0071
DBN	0.0035	14.77	0.0041
LHRSS-ELM	0.0055	14.92	0.0056
SDAE-NN	0.0031	23.54	0.0034
LHSS-HELM	**0.0029**	15.42	**0.0032**

As another comparison, the prediction results and MSE errors of ELM, SS-ELM, HELM, and LHSS-ELM are given in Fig. 8.2. The same parameter set was used in the model training, and 500 labeled samples and 1000 unlabeled samples were used for the entire semi-supervised learning process. LHSS-HELM presents an intuitive improvement in

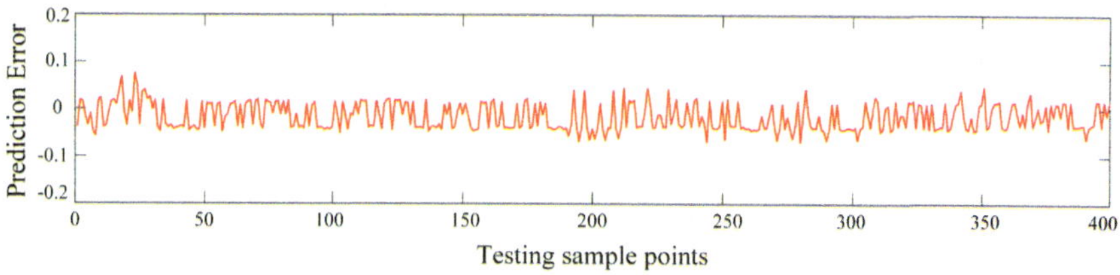

(a) The proposed method

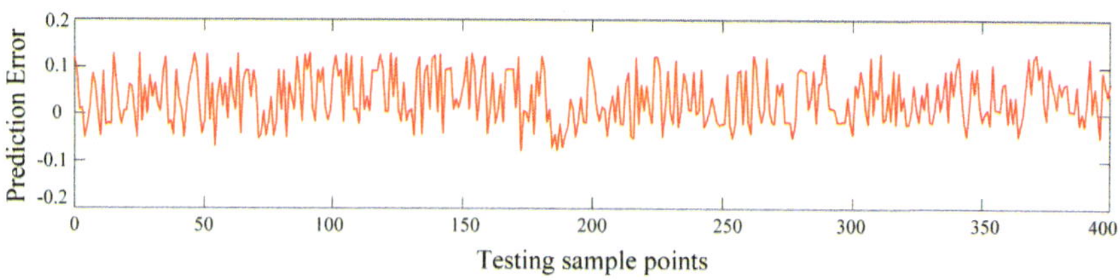

(b) HELM

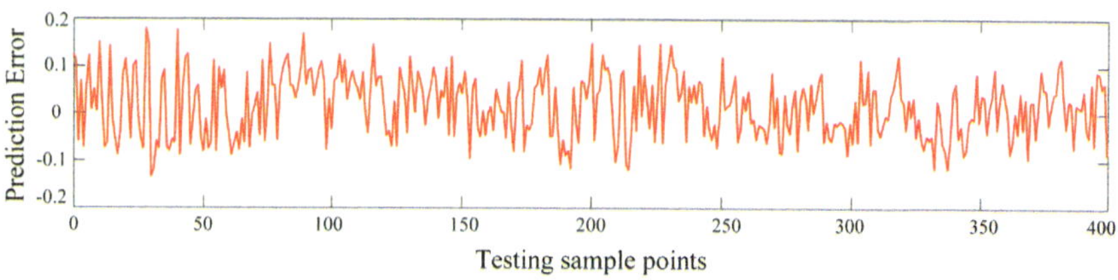

(c) SS-ELM

Figure 8.2 Testing error comparison with various numbers of unlabeled samples for various algorithms.

terms of prediction performance. Furthermore, the comparison of four existing soft sensor methods, namely, semi-supervised probabilistic principal component regression (SS-PPCR), Co-PLS, deep belief network (DBN), and denoising autoencoder with a neural network (SDAE-NN), and LHSS-HELM. Under the same training conditions, soft sensor models were built. Fig. 8.3 shows the comparison results. The SS-PPCR and Co-PLS are semi-supervised approaches that can fully use unlabeled data for model training. However, the inherent linearity considerably limits the enhancement of prediction performance. The DBN and SDAE-NN are derived from DL domains, which have been implemented in some quality variable regression domains. The DBN method is pre-trained by a restricted Boltzmann machine, and then the parameter for initializing BP-NN is obtained. For SDAE-NN, the same method is introduced to

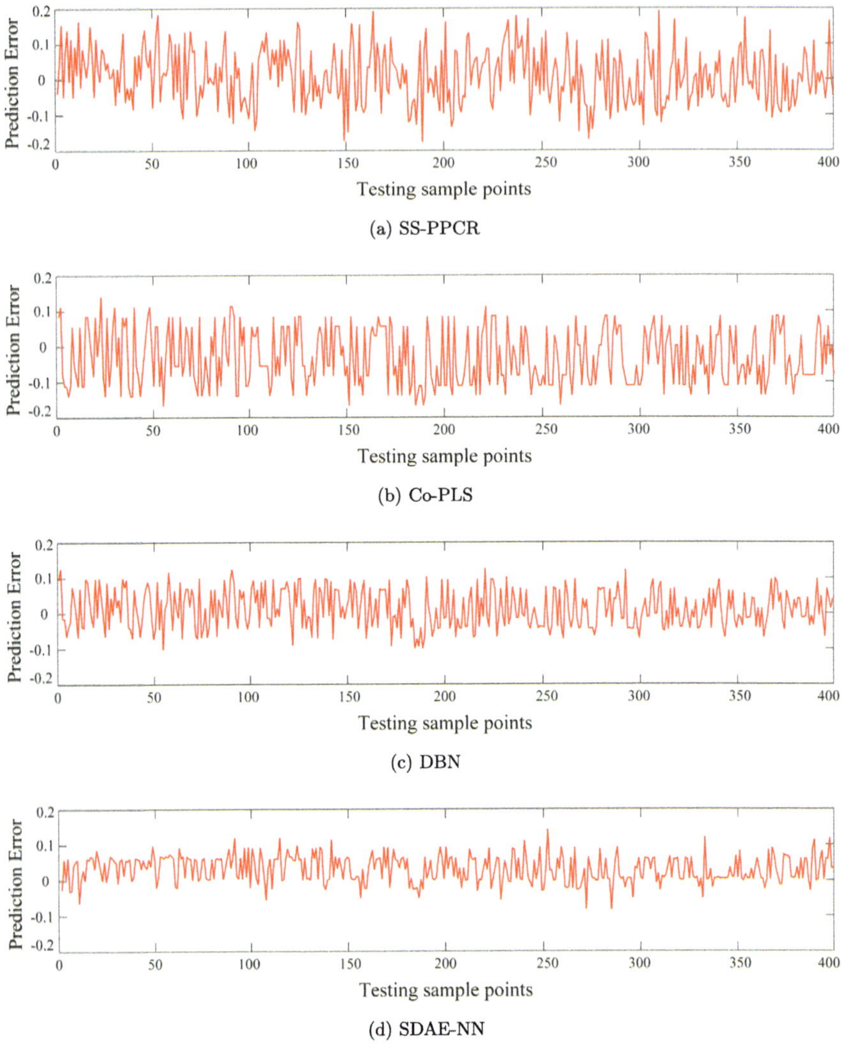

Figure 8.3 Testing error comparison with various numbers of unlabeled samples for various algorithms.

obtain the initial weights of NN. For a fair comparison, the number of hidden neurons and hidden layer are the same as with LHSS-HELM. The results demonstrate that both DBN and SDAE-NN perform excellently in quality prediction, indicating that the deep architecture extracts superior and complex features in the high layer. The training speed in DBN and SDAE-NN is slow because the BP algorithm is used in model training.

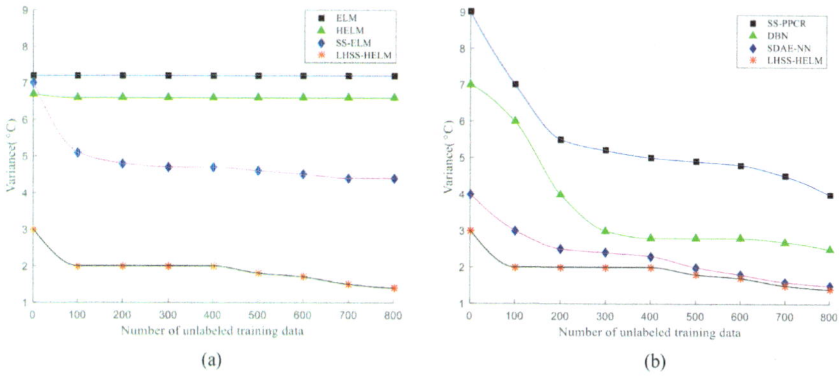

Figure 8.4 Testing variance comparison with various numbers of unlabeled samples for various algorithms: (a) ELM-based comparison and (b) other comparison methods.

The results of all the data showed the mean values of 10 iterations of experiments. The variances with various numbers of unlabeled examples are described in Fig. 8.4. The results show that the LHSS-HELM has lower variance and superior robustness compared to other existing soft sensor methods.

8.5 Conclusions

In this chapter, the DL based on HAE and ELM were used to build a semi-supervised soft sensor modeling. Traditional semi-supervised ELM lacks extrapolating power and robustness. To solve this problem, we proposed a weighted Laplacian–Hessian integrated hierarchical SS-ELM algorithm and employed sensor modeling. First, HAEs were used for deep complex feature extraction with all labeled and unlabeled samples. Then a semi-supervised ELM based on Laplacian–Hessian regularization was used for quality variable regression. To obtain a superior LHSS-HELM model, we proposed a pair of trade-off parameter sequences to update the hyper-parameters. After combining multi-layer unsupervised feature extraction and semi-supervised regression, the proposed LHSS-HELM considerably improved the extrapolating ability, performance, and robustness. The soft model for CT prediction implemented in the industrial aluminum reduction cell demonstrated that the performance outperforms the existing state-of-the-art soft sensor models. The extension in LHSS-HELM can be further applied to multi-class classification or regression problems and give new sights for these domains.

References

[1] Xiaofeng Yuan, Biao Huang, Yalin Wang, Chunhua Yang, Weihua Gui, Deep learning-based feature representation and its application for soft sensor modeling with variable-wise weighted SAE, IEEE Transactions on Industrial Informatics 14 (7) (2018) 3235–3243.

[2] Xiaoxia Chen, Ting Lan, Xuhua Shi, Chudong Tong, A semi-supervised linear–nonlinear least-square learning network for prediction of carbon efficiency in iron ore sintering process, Control Engineering Practice 100 (2020) 104454.

[3] Xiao Wang, Han Liu, Data supplement for a soft sensor using a new generative model based on a variational autoencoder and Wasserstein GAN, Journal of Process Control 85 (2020) 91–99.

[4] Kai Jiang, Shizhu Shi, Hosein Moazanigoodarzi, Chuan Hu, Souvik Pal, Fengjun Yan, Temperature distribution estimation via data-driven model and adaptive Kalman filter in modular data centers, Proceedings of the Institution of Mechanical Engineers. Part I, Journal of Systems and Control Engineering 234 (7) (2020) 809–819.

[5] Daoguang Yang, Hamid Reza Karimi, Kangkang Sun, Residual wide-kernel deep convolutional auto-encoder for intelligent rotating machinery fault diagnosis with limited samples, Neural Networks 141 (2021) 133–144.

[6] J. Wang, J. Liang, J. Cui, et al., Semi-supervised learning with mixed-order graph convolutional networks, Information Sciences 573 (2021) 171–181.

[7] Y.L. He, Q. Hua, Q.X. Zhu, et al., Enhanced virtual sample generation based on manifold features: applications to developing soft sensor using small data, ISA Transactions 126 (2022) 398–406, https://doi.org/10.1016/j.isatra.2021.07.033.

[8] Z. Chen, Z. Ge, Knowledge automation through graph mining, convolution and explanation framework: a soft sensor practice, IEEE Transactions on Industrial Informatics 18 (9) (2022) 6068–6078, https://doi.org/10.1109/TII.2021.3127204.

[9] D. Kolar, D. Lisjak, M. Pajak, et al., Fault diagnosis of rotary machines using deep convolutional neural network with wide three axis vibration signal input, Sensors 20 (14) (2020) 4017.

[10] S. Xie, Y. Xie, T. Huang, et al., Generalized predictive control for industrial processes based on neuron adaptive splitting and merging RBF neural network, IEEE Transactions on Industrial Electronics 66 (2) (2019) 1192–1202.

[11] Y. Li, Hamid Reza Karimi, Q. Zhang, et al., Fault detection for linear discrete time-varying systems subject to random sensor delay: a Riccati equation approach, IEEE Transactions on Circuits and Systems. I, Regular Papers 65 (2018) 1707–1716.

[12] Guang-Bin Huang, Qin-Yu Zhu, Chee-Kheong Siew, Extreme learning machine: theory and applications, Neurocomputing 70 (1–3) (2006) 489–501.

[13] Yongxiang Lei, Xiaofang Chen, Ming Min, et al., A semi-supervised Laplacian extreme learning machine and feature fusion with CNN for industrial superheat identification, Neurocomputing 381 (2020) 186–195.

[14] Yongxiang Lei, Xiaofang Chen, Yongfang Xie, An improved cell situation identification approach with convolutional neural network and wavelet extreme learning machine, Proceedings of the Institution of Mechanical Engineers. Part I, Journal of Systems and Control Engineering 235 (10) (2021) 1898–1905.

[15] Y. Jiang, P. Li, Y. Lei, A novel superheat identification of aluminum electrolysis with kernel semi-supervised extreme learning machine, Journal of Physics. Conference Series 1631 (1) (2020) 012005.

[16] M. Min, X. Chen, Y. Lei, et al., A novel kernel-based extreme learning machine with incremental hidden layer nodes, IFAC-PapersOnLine 53 (2) (2020) 11836–11841.

[17] J. Ke, C. Gong, T. Liu, et al., Laplacian Welsch regularization for robust semisupervised learning, IEEE Transactions on Cybernetics 52 (2022) 164–177.

[18] Bingbing Shen, Le Yao, Zhiqiang Ge, Nonlinear probabilistic latent variable regression models for soft sensor application: from shallow to deep structure, Control Engineering Practice 94 (2020) 104198.
[19] P. Sussner, I. Campiotti, Extreme learning machine for a new hybrid morphological/linear perceptron, Neural Networks 123 (2020) 288–298.
[20] Bruno Leonardo Soares Da Silva, Fábio Kenji Inaba, E.O.T. Salles, et al., Outlier robust extreme machine learning for multi-target regression, Expert Systems with Applications 140 (2020) 112877.
[21] Y. Liao, R. Huang, J. Li, et al., Deep semisupervised domain generalization network for rotary machinery fault diagnosis under variable speed, IEEE Transactions on Instrumentation and Measurement 69 (10) (2020) 8064–8075.
[22] S. Ying, Z. Wen, J. Shi, et al., Manifold preserving: an intrinsic approach for semisupervised distance metric learning, IEEE Transactions on Neural Networks and Learning Systems 29 (7) (2017) 2731–2742.
[23] Gao Huang, Shiji Song, Jatinder N.D. Gupta, Cheng Wu, Semi-supervised and unsupervised extreme learning machines, IEEE Transactions on Cybernetics 44 (12) (2014) 2405–2417.
[24] C.M. Wong, C.M. Vong, P.K. Wong, et al., Kernel-based multilayer extreme learning machines for representation learning, IEEE Transactions on Neural Networks and Learning Systems 29 (3) (2018) 757–762.
[25] S. Sankaranarayanan, N. Sivakumaran, T.K. Radhakrishnan, et al., Dynamic soft sensor based parameters and demand curve estimation for water distribution system: theoretical and experimental cross validation, Control Engineering Practice 102 (2020) 104544.
[26] S. Fu, W. Liu, D. Tao, et al., HesGCN: Hessian graph convolutional networks for semi-supervised classification, Information Sciences 514 (2020) 484–498.
[27] X. Yuan, Y. Wang, C. Yang, et al., Weighted linear dynamic system for feature representation and soft sensor application in nonlinear dynamic industrial processes, IEEE Transactions on Industrial Electronics 65 (2) (2018) 1508–1517.
[28] J. Loy-Benitez, S.K. Heo, C.K. Yoo, Soft sensor validation for monitoring and resilient control of sequential subway indoor air quality through memory-gated recurrent neural networks-based autoencoders, Control Engineering Practice 97 (2020) 104330.
[29] J. Zhang, Y. Wan, Z. Chen, et al., Non-negative and local sparse coding based on l_p-norm and Hessian regularization, Information Sciences 486 (2019) 88–100.
[30] Y. Lei, L. Cen, X. Chen, et al., A hybrid regularization semi-supervised extreme learning machine method and its application, IEEE Access 7 (2019) 30102–30111.
[31] Ping Wang, Yichao Yin, Xiaogang Deng, Yingchun Bo, Weiming Shao, Semi-supervised echo state network with temporal–spatial graph regularization for dynamic soft sensor modeling of industrial processes, ISA Transactions 130 (2022) 306–315.

CHAPTER 9

A self-supervised prediction framework based on deep long short-term memory for aluminum electrolysis: SSDLSTM

9.1 Introduction

The difficulty of timely measurement of pot temperature (PT) in aluminum electrolysis reduction has been a challenging problem beyond technological solutions. The commonly strategies are physical devices and off-line lab analyzers. However, there are many factors that should be taken into consideration, such as time delay, severe environment, expensive artificial and device cost; for instance, there are lots of data-driven models that have some related limitations [1,2]. Moreover, for the stable and highly efficient production, the PT measurement by the worker with physical device cannot meet the performance requirements for key-variable prediction.

To alleviate the above-mentioned problems, some techniques have been proposed in the last decades. For example, Lei et al. [3–5] proposed some extreme learning machine methods with high computational efficiency for training the single hidden layer network. Zeng et al. developed a support vector machine for the cell superheat degree classification [6,7]. However, the collection of labeled data in industrial aluminum production is intractable and time-consuming. The scarcity of the samples greatly reduce the generation ability of the existing models. To learn the time initial features, some time-sequential methods are also widely used in the industrial process. For example, Sun et al. proposed a deep LSTM method for the speech recognition enhancement, which has improved the performance of the traditional recurrent neural networks (RNNs). Cong et al. proposed a dynamic robust method for the output feedback attacks application [7]. In the past, RNN has been incredibly applied to a variety of problems, for instance, speech recognition, language modeling, translation, image captioning, etc. However, training an RNN-based model is a hard task because the optimal goal is a time-dependent sequence function, which

Learning-Based Soft Sensing and Predictions for Process Industries
https://doi.org/10.1016/B978-0-44-336759-5.00018-4

relies on all the historical data, and the long sequence reliability, and the gradient exploding problem is very apparent, especially when the network is very large.

Recently, data-driven measurement methods have gained a lot of popularity. Typical data-driven soft modeling methods include principal component analysis (PCA), linear graph embedding (LGE), extreme learning machine (ELM), support vector machine (SVM), artificial neural network (ANN), etc. The core principle is to learn the nonlinear projection between the input and output. A critical condition for guaranteeing the desired performance of these methods is that it needs the abundant labeled samples to train the large-scale network architectures. Further, the gradient descent and exploding problems are obvious, especially when the network is very large, which greatly limits the performance of these deep architectures for large-scale systems. Indeed, deep learning unit has been successfully applied to many areas such as traffic flow prediction, speech recognition, and computer vision. RNN is a popular sequence model, which has been widely applied to describe temporal dynamic behavior for time sequence such as language sequence tasks. For example, Yuan et al. [8] used the RNN to predict and evaluate the degree-of-cure in a fiber composites system. However, standard RNN-based models are often difficult to train and have the difficulty in modeling long sequences. Long-short-term memory (LSTM) is a variant of RNNs proposed by Hochreiter et al. [9], which is capable of learning long-term dependencies. Instead of using the hidden state from the previous time step and the current input in a single *tanh* layer to implement recurrence [10,1], LSTMs use memory cells and nonlinear gates, which interact in a very specific way. LSTM model is relatively less used in industrial domains. In the basic LSTM unit, the activation function of Sigmoid cannot be replaced by other functions because it learns the memory gate and projects the hidden unit to the range from 0 to 1. The tanh activation function can be used as another alternative.

Due to the traditional supervised and reinforcement learning requirements, the manual disturbance and its performance will rapidly decrease when there are not abundant labeled samples. Self-supervised learning (SSL) has attracted promising attention in recent years. Some classical self-supervised learning models have improved the few-shot learning accuracy (e.g., MoCo [11], SimCLR [12], and BYOL [13]), which has reduced the gap between the supervised learning methods. SSL, proposed in [14], automatically generates the feature and label representation for the input data. It creates the supervision and labels through pre-used model and often used interchangeably with unsupervised learning. The self-supervised learning methods can improve the accuracy of the model while not mainly relying on the annotated samples. The uncurated and unlabeled data are used in the model to learn the information of the samples. The traditional method of the pot temperature is based on the supervised learning; however, the aluminum faces the limitation of the labeled data, and the supervised model cannot be implemented in such a constraint of labeled data. According to the mechanical analysis, there are many underlying manifold distributions from various different quality variables. Self-supervised learning is in its infant period, and it learns the information and features from its inner geometry or manifold. Further, in the practical industrial process, there is a sequent link for the pot temperature and other variables. Based on the above reasons, we proposed the novel model of self-supervised LSTM and first apply it in the aluminum industry.

Based on the aforementioned discussion, we propose a self-supervised deep model based on LSTM (SSDLSTM) for PT prediction in aluminum production. First, the input images are selected according to the expert experience. Second, the training samples are collected for training the whole self-supervised LSTM model. Finally, the on-line model is used for PT prediction. It is worth highlighting the following contributions.

- An enhanced LSTM architecture is proposed such that the target output is extracted from the learning process. The traditional LSTM learns the parameters in an unsupervised learning framework, which cannot fully use the target information. Extracting and processing of meaningful information from industry are a challenging and valuable task.
- Based on the proposed LSTM unit, a soft model is further constructed with long-short-term memory network for PT prediction. At each sampling instant, the quality is exploited into the unsupervised LSTM to learn the dynamic latent hidden state, which can largely keep the relevancy of input historical data and current sampling instant.
- Comprehensive experimental studies have been conducted to evaluate the performance and generalization. Compared with RNNs and other competing algorithms, the proposed method can realize PT detection on-line. It is shown that the accuracy of our proposed method is superior to the existing techniques, such as LSTM and RNN. The dynamic performance also satisfied the industrial requirements in aluminum electrolysis production.

9.2 Problem statement

In this part, we use the proposed self-supervised LSTM to construct a soft modeling for industrial aluminum electrolysis reduction. The model was developed as self-supervised LSTM for the PT process prediction, and the approach frame is presented in the equation

$$\hat{y}(t) = \sigma\left(Vh_{(t)} + b\right), \tag{9.1}$$

where $y(t)$ is the predicted value of PT variable. Eq. (9.1) is a basic and classic prediction unit for the SSDLSM model, $y(t)$ is the prediction of pot temperature at the instant time t, which is the whole model output of SSDLSTM, $(Vh(t) + b)$ is the uniformed LSTM output toward the softmax activation function, and σ is the activation, which commonly is a sigmoid function, which means that the gate output ranges from 0 to 1. This equation can be automatically updated from the instant t to the instant $t+1$, and historical information can be reserved from the LSTM unit.

As stated in [5], the whole training process uses the BPTT algorithm to learn the network parameters. The proposed block learns the dynamic sequential features from the historical data, and the predicted PT value can be used to continuously update the network hidden parameters.

9.3 Main framework

In this part, we present the formulation of the proposed LSTM-based prediction framework. As given in Fig. 9.1, the original flame data was split into several pieces of training samples. The relative sampling frequency is 5 seconds for one interval. The samples are fully connected with high-dimensional feature input, and a stacked LSTM learns the temporal information about the aluminum electrolysis process. In the proposed SSDLSTM framework, the KL divergence is used to compute the similarity in the training process. Since the label y_i is missing, the hidden features are constructed by the pair (z_i, y_i). Then the log-likelihood function can be written as the following constraint:

$$\begin{aligned} \max \quad & \mathrm{L}^s\left(\boldsymbol{x}_i, \boldsymbol{y}_i\right) = \\ & \mathrm{E}_{q(z)}\left[\ln p\left(\boldsymbol{x}_i, \boldsymbol{y}_i \mid \boldsymbol{z}\right)\right] - KL[q(\boldsymbol{z}) \| p(\boldsymbol{z})], \end{aligned} \tag{9.2}$$

where $\mathrm{E}_q[\cdot]$ is the expectation of the distribution p, and $KL[q(z)\|p(z)] = \int q(z) \ln \frac{q(z)}{p(z)} dz$ is the Kullback–Leibler (KL) divergence to evaluate the

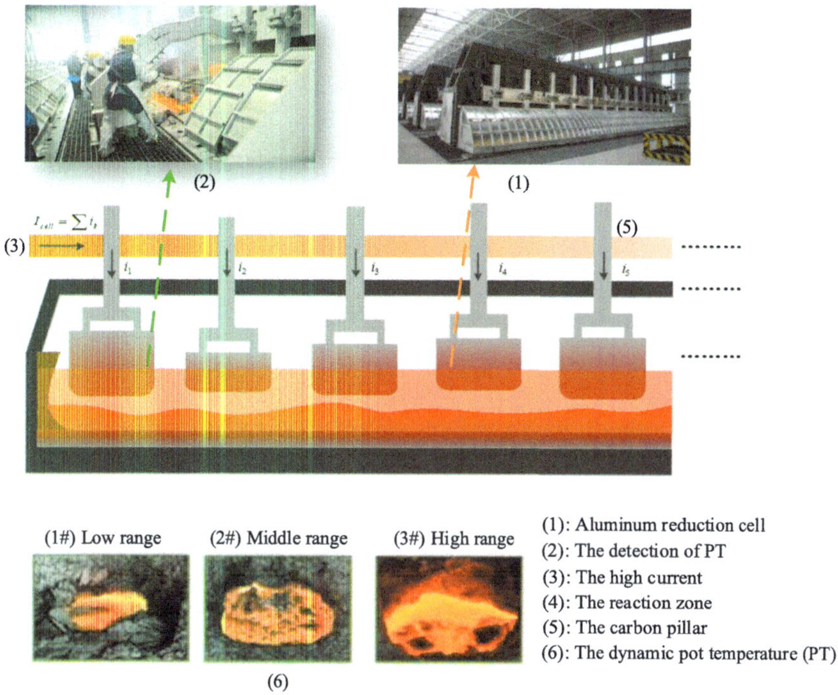

Figure 9.1 Structure diagram of aluminum reduction cell.

dissimilarity between the two labels. Consider the following conditional distribution of x, y, and z:

$$p(z, y \mid x) = p(z \mid y, x) p(y \mid x), \tag{9.3}$$

$$p(\boldsymbol{x} \mid \mathbf{z}, \boldsymbol{y}) = p(\boldsymbol{x} \mid \boldsymbol{z}). \tag{9.4}$$

The static probability is derived as follows:

$$\begin{aligned}
&\mathrm{L}^{u}\left(\boldsymbol{x}_{i}^{u}\right) = \\
&\mathrm{E}_{p(\boldsymbol{y}\mid\boldsymbol{x}_{i}^{u})}\left[\mathrm{E}_{p(\boldsymbol{z}\mid\boldsymbol{x}_{i}^{u},\boldsymbol{y})}\left[\ln p\left(\boldsymbol{x}_{i}^{u}\mid\boldsymbol{z}\right)\right]\right] + \mathrm{H}\left(p\left(\boldsymbol{y}\mid\boldsymbol{x}_{i}^{u}\right)\right) \\
&-\mathrm{E}_{p(\boldsymbol{y}\mid\boldsymbol{x}_{i}^{u})}\left[KL\left[p\left(\boldsymbol{z}\mid\boldsymbol{y},\boldsymbol{x}_{i}^{u}\right)\|p(\boldsymbol{z})\right]\right] \\
&+\mathrm{E}_{p(\boldsymbol{y}\mid\boldsymbol{x}_{i}^{u})}\left[\mathrm{E}_{p(\boldsymbol{z}\mid\boldsymbol{y},\boldsymbol{x}_{i}^{u})}\left[\ln p(\boldsymbol{y}\mid\boldsymbol{z})\right]\right] \\
&= \mathrm{E}_{p(\boldsymbol{y}\mid\boldsymbol{x}_{i}^{u})}\left[\mathrm{L}^{s}\left(\boldsymbol{x}_{i}^{u},\boldsymbol{y}\right)\right] + \mathrm{H}\left(p\left(\boldsymbol{y}\mid\boldsymbol{x}_{i}^{u}\right)\right).
\end{aligned} \tag{9.5}$$

By introducing the interval term the above equation can be expressed as follows:

$$
\begin{aligned}
&KL\left[p\left(z, y \mid x_i^u\right) \| p(z, y)\right] \\
&= \iint p\left(y \mid x_i^u\right) p\left(z \mid y, x_i^u\right) \ln \tfrac{p(z \mid y, x_i^u) p(y \mid x_i^u)}{p(z) p(y \mid z)} dz dy \\
&= \iint p\left(y \mid x_i^u\right) p\left(z \mid y, x_i^u\right) \ln \tfrac{p(z \mid y, x_i^u)}{p(z)} dz dy \\
&+ \iint p\left(y \mid x_i^u\right) p\left(z \mid y, x_i^u\right) \ln p\left(y \mid x_i^u\right) dz dy \\
&= \iint p\left(y \mid x_i^u\right) p\left(z \mid y, x_i^u\right) \ln p(y \mid z) dz dy \\
&= \mathrm{E}_p\left(y \mid x_i^u\right)\left[KL\left[p\left(z \mid y, x_i^u\right) \| p(z)\right]\right] - \mathrm{H}\left(p\left(y \mid x_i^u\right)\right) \\
&-\mathrm{E}_{p(y \mid x_i^u)}\left[\mathrm{E}_{p(z \mid y, x_i^u)}[\ln p(y \mid z)]\right],
\end{aligned}
\tag{9.6}
$$

where z denotes the hidden variables, and E_p is the feature expectation of the output variables under the constraint of input $\mathbf{x}$. As shown in Fig. 9.2,

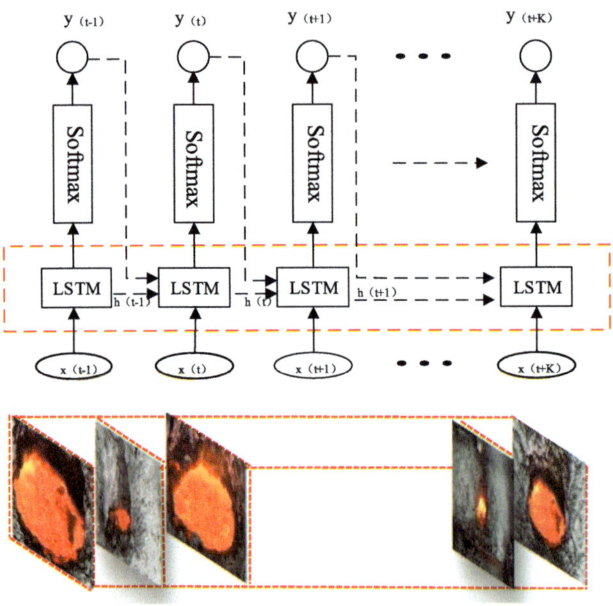

Figure 9.2 SSDLSTM network structure.

for an input image X_r resized to 384∗384, full-connected neurons project the input samples into the long-term sequence units. The procedure of the proposed SS-DLSTM is described in the following steps:

1. Initialize the structure of the proposed SSDLSTM, which covers the number of the LSTM unit and the T time sequences;

2. Collect the approximate input quality images X_t in the aluminum electrolysis process;
3. Set the learning rate, iteration epoch, bath-size, etc;
4. Train the model with historical data sample in a self-supervised circle, update the next t time sequence pot temperature prediction with Eqs. (9.3)–(9.6);
5. Test the performance of the proposed SSDLSTM model with the available unlabeled samples and compute the accuracy and error $RMSE$, R^2;
6. Use the proposed model for further temporal pot temperature prediction.

Remark 1. In the proposed algorithm, a self-supervised SSDLSTM approach is developed, and KL divergence is randomness. SSDLSTM learns the time instant features and avoids reducing the performance under the scarcity of the labels. The aluminum electrolysis production is a traditional case to verify the effectiveness of the proposed method. Considering the multiplication and addition operations in the LSTM prediction output unit, the computation complexity of the proposed SSDLSTM is $O(N^2)$.

In this part, we use the proposed self-supervised LSTM to construct a soft modeling for industrial aluminum electrolysis reduction. The model was developed as self-supervised LSTM for the PT process prediction, and the approach frame is presented in the equation

$$\hat{y}(t) = \sigma\left(Vh_{(t)} + b\right), \tag{9.7}$$

where $y(t)$ is the predicted value of PT variable. Eq. (9.7) is a basic and classic prediction unit for the SSDLSM model, where $y(t)$ is the prediction of pot temperature at the time instant t, which is the whole model output of SSDLSTM, $(Vh(t) + b)$ is the uniformed LSTM output toward the softmax activation function, and σ is the activation, which commonly is a sigmoid function, which means that the gate output ranges from 0 to 1. This equation can be automatically updated from the instant t to the instant $t+1$, and historical information can be reserved from the LSTM unit.

As stated in [5], the whole training process uses the BPTT algorithm to learn the network parameters. The proposed block learns the dynamic sequential features from the historical data, and the predicted PT value can be used to continuously update the network hidden parameters.

9.4 Simulation results

The SD situation index is the temporal time prediction for the proposed LSTM architecture. The corresponding results are given in Table 9.1. Since LSTM uses a cell memory to store the real-time status information, the performance of LSTM outperforms RNN. The training and testing RMSE of LSTM are 0.0246 and 0.0513, respectively, which are much lower than those of RNN. The other self-supervised methods, such as MoCo and ARTMAP, are also compared in our experiments. The results show that SSDLSTM has the competitive performance with the supervised methods, see Tables 9.2, 9.3.

Table 9.1 Performance comparison for RNN, SS-LSTM, CNN, and the proposed methods. Trtime: Training time (s), Var: Variance.

Methods	LSTM	RNN	Attention-LSTM	CNN	Ours
RMSE	0.269	0.569	0.348	0.208	0.215
R^2	0.062	0.083	0.012	0.021	**0.008**

Table 9.2 Comparative experimental results of supervised semi-supervised learning.

Form	Supervised		Semi-supervised	
Temperature zone	upper	lower	upper	lower
RMSE	0.4838	0.5567	0.6564	0.8843
R^2	−0.4001	−0.3430	0.2612	0.3218

Table 9.3 Training times and variance for different algorithms. A: low-temperature zone, B: middle-temperature zone, C: upper-temperature zone.

Categories	Te_acc			Tr_acc			Tr_time(s)		
	A	B	C	A	B	C	A	B	C
Supervised									
DNN	0.504	0.618	2.626	0.456	0.621	0.481	24.76	24.77	24.77
SimCLR[12]	0.362	1.023	0.128	0.523	0.722	0.813	4.73	5.33	5.22
BYOL[13]	0.233	0.732	0.625	0.616	0.714	**0.881**	6.88	7.31	7.37
Semi-supervised									
SWPPLSR[6]	0.342	0.534	0.636	0.324	0.566	0.698	3.71	3.45	4.23
CNN-LapsELM[7]	0.645	0.423	0.245	0.601	0.711	0.748	3.42	3.44	3.28
Self-supervised									
MoCo[11]	0.674	0.828	0.867	0.086	0.785	0.761	6.21	6.20	6.52
ARTMAP[6]	0.562	0.623	0.648	0.456	0.725	0.737	7.93	8.40	7.22
Ours	0.693	0.832	0.845	0.656	**0.895**	0.841	2.59	**2.34**	2.53

In the proposed SSDLSTM approach, the RMSE of the training and testing values are 0.0247 and 0.0078, respectively, so the prediction accuracy is better than those of other competitive methods. Under the constraint of the limited number of samples, the SSDLSTM maintains its accuracy and generation ability. The training accuracy is up to 0.895 in the middle-temperature zone, it outperforms other supervised and semi-supervised methods.

To give a promising and competitive comparison, the testing RMSE and training RMSE are both calculated in the process of the iterations. Fig. 9.3

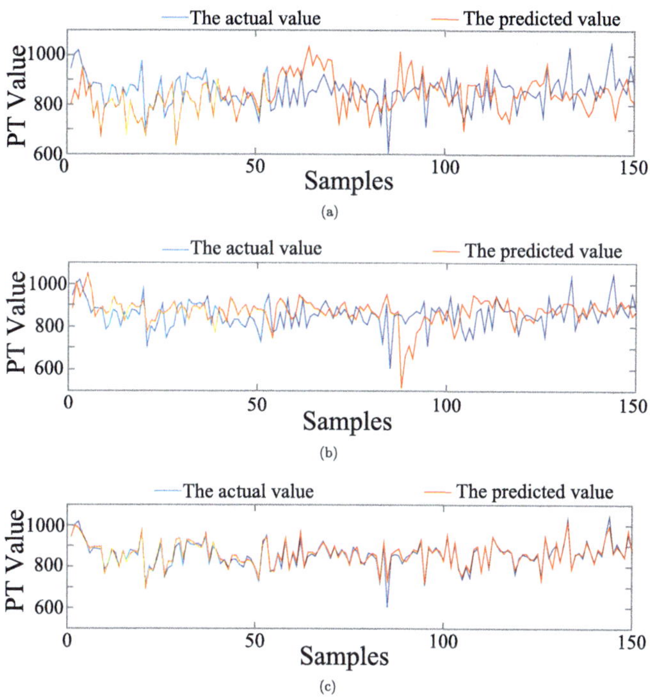

Figure 9.3 Testing error comparison with the different numbers of unlabeled samples for different algorithms: (a) RNN; (b) LSTM; (c) SSDLSTM.

indicates the results of the related experiments in the industrial aluminum electrolysis. From these results, the proposed SSDLSTM outperforms the traditional RNN and CNN, and furthermore, with the increase of the number of iteration epoch, the RMSE rapidly decreased. Under the circumstance of the large number of training data, the LSTM can gain and preserve the long-term input information, so the temporal prediction is better than that of RNN. The results are perfectly suited for learning

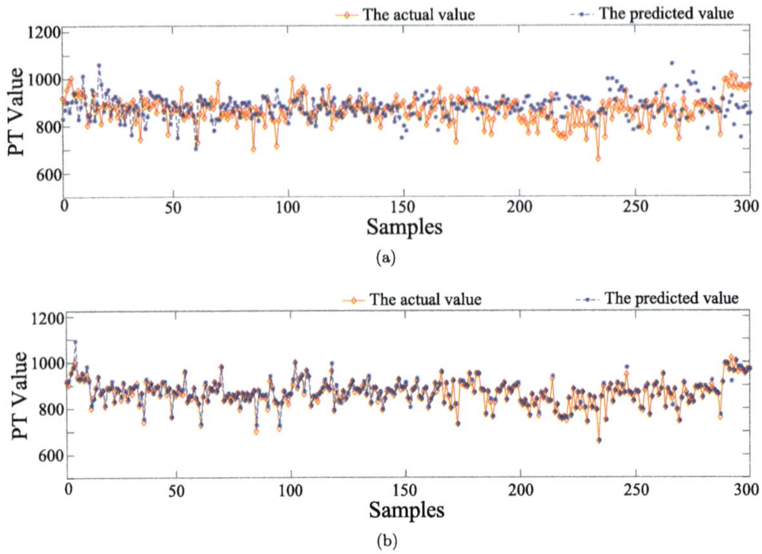

Figure 9.4 Testing error comparison with various numbers of unlabeled samples for algorithms (a) Attention-LSTM and (b) SSDLSTM.

the features and PT prediction in the aluminum electrolysis analysis. The established model has crucial independence properties with a feasible solution, see Fig. 9.4.

9.5 Conclusions

In this framework, we propose a novel self-supervised LSTM to improve the long-time dependency of RNN. The traditional LSTM is essentially an unsupervised learning method. Then a deep LSTM is further built based on the proposed LSTM unit. The soft modeling and application in industrial aluminum demonstrated that the performance and generalization ability is competitive with the RNN and LSTM methods. The categories of supervised and semi-supervised learning are both evaluated in the experimental comparison. The proposed SSDLSTM can realize the dynamics and nonlinear real-time prediction of PT and avoid the time-consuming and expensive data annotations. For further sight, the self-supervised deep LSTM should consider the effect of the noise and outlier, and the proposed technique has a great generation ability, which can be further applied to many other domains.

References

[1] Yu Guan, Thomas Piltz, Ensembles of deep LSTM learners for activity recognition using wearables, Proceedings of the ACM on Interactive, Mobile, Wearable and Ubiquitous Technologies 1 (2) (2017) 1–28.
[2] Alaa Sagheer, Mostafa Kotb, Time series forecasting of petroleum production using deep LSTM recurrent networks, Neurocomputing 323 (2019) 203–213.
[3] Yongxiang Lei, Xiaofang Chen, Ming Min, et al., A semi-supervised Laplacian extreme learning machine and feature fusion with CNN for industrial superheat identification, Neurocomputing 381 (2020) 186–195.
[4] Jongmin Ryu, Min-Ho Yang, Joonhwan Lim, Unsupervised feature learning for self-tuning neural networks, Neural Networks 133 (2021) 103–111.
[5] Y. Lei, L. Cen, X. Chen, et al., A hybrid regularization semi-supervised extreme learning machine method and its application, IEEE Access 7 (2019) 30102–30111.
[6] H.J. Yoo, R.L. Pimmel, Short term load forecasting using a self-supervised adaptive neural network, IEEE Transactions on Power Systems 14 (2) (1999) 779–784.
[7] Cong Huang, Serdar Coskun, Jun Wang, Peng Mei, Quan Shi, Robust H_∞ dynamic output-feedback for CACC with ROSSs subject to RODAs, IEEE Transactions on Vehicular Technology 71 (2022) 137–147, Volume inferred from publication year.
[8] Xiaofeng Yuan, Lin Li, Yalin Wang, Nonlinear dynamic soft sensor modeling with supervised long short-term memory network, IEEE Transactions on Industrial Informatics 16 (5) (2019) 3168–3176.
[9] Sepp Hochreiter, Jürgen Schmidhuber, Long short-term memory, Neural Computation 9 (8) (1997) 1735–1780.
[10] Wentao Zhu, Cuiling Lan, Junliang Xing, et al., Co-occurrence feature learning for skeleton based action recognition using regularized deep LSTM networks, arXiv preprint, arXiv:1603.07772, 2016.
[11] Kaiming He, Haoqi Fan, Yuxin Wu, Saining Xie, Ross Girshick, Momentum contrast for unsupervised visual representation learning, in: Proceedings of the IEEE/CVF Conference on Computer Vision and Pattern Recognition, 2020, pp. 9729–9738.
[12] Ting Chen, Simon Kornblith, Mohammad Norouzi, Geoffrey Hinton, A simple framework for contrastive learning of visual representations, in: International Conference on Machine Learning, PmLR, 2020, pp. 1597–1607.
[13] Jean-Bastien Grill, Florian Strub, Florent Altché, Corentin Tallec, Pierre Richemond, Elena Buchatskaya, Carl Doersch, Bernardo Avila Pires, Zhaohan Guo, Mohammad Gheshlaghi Azar, et al., Bootstrap your own latent: a new approach to self-supervised learning, Advances in Neural Information Processing Systems 33 (2020) 21271–21284.
[14] Ashish Jaiswal, Ashwin Ramesh Babu, Mohammad Zaki Zadeh, Debapriya Banerjee, Fillia Makedon, A survey on contrastive self-supervised learning, Technologies 9 (1) (2020) 2.

CHAPTER 10

A self-supervised prediction framework based on convolutional deep long short-term memory for aluminum temperature application: CNN-SSDLSTM

10.1 Introduction

With the advent of artificial intelligence and information technologies, the productivity of the modern industrial system continuously increases [1,2]. In aluminum electrolysis production, the superheat degree is an important index for the indicator and monitoring of the whole production situation, lengthening the lifespan of the reduction cell, and improving the current efficiency [3–6]. However, cell temperature detection suffers from many limitations, such as the hard environment, expensive device cost, and large delay. In the aluminum reduction reaction, some catalysts can be used to decrease the crystal temperature of the electrolysis, and so the current efficiency has also been improved. Recently, some data-driven methods have been proposed to detect the temperature situation [7,8]. The main strategy can be summarized by the following three categories: 1) the model-centric model, also the "white" model; an obvious index is that the features and parameters can be visualized in the learning process. The model-based method needs to know the accurate dynamic mathematical model of the specific plant. Some works have also been reported about this field, for example, [9]. However, to sum up, the performance of those models largely relies on the real-time mathematical model, and their generation ability is limited. Furthermore, in some industry scenarios, it is hard to obtain some mathematical model, and thus the generation ability of these models is greatly limited. 2) The pair-wise data-centric intelligent models, also called "black" models, which use the deep architecture to project the input and

Learning-Based Soft Sensing and Predictions for Process Industries
https://doi.org/10.1016/B978-0-44-336759-5.00019-6

output between the learning samples, have been proposed in the literature [10,11].

Recently, deep learning and intelligent algorithms played an important role in industrial fields. With the abundant training data, the deep architecture can learn a good nonlinear projection of the data distribution, and thus a remarkable performance is achieved. Some classification and regression tasks have been fully implemented in the related applications [1–12]. However, because the known models are primarily based on the supervised learning method, their performance greatly suffer when limited to labeled data. Although some semi-supervised learning methods have been proposed to solve this issue, the semi-supervised models rely on a few samples, and if these labeled samples have bad quality, then the model has low robustness and performance. Furthermore, in the deep architecture, gradient descent and gradient explosion are substantial in training the model, resulting in over-fitting.

In the recent several years, self-supervised learning has been the hottest topic in the sub-domains of machine learning [13–16]. In SSL a learning machine captures the dependencies between input variables, some of which may be observed, denoted $x_1, \ldots, x_N$, and others are not always observed, denoted $\{x_{N+1}, \ldots, x_{N+s}\}$. SSL pre-training has revolutionized natural language processing and makes rapid progress in speech and image recognition [17–20]. SSL may enable machines to learn predictive models of the world through observation and to learn representations of the perceptual world, thereby reducing the number of labeled samples or rewarded trials to learn a downstream task [14]. The core principle, resulting in the development of self-supervised learning, is that under the constraint and limitation of the labeled data, how can we build some model autonomously to learn its feature and knowledge? The self-supervised learning is mainly classified into three categories: the basic self-supervised, contrastive model, and learning models. The advantage of the proposed model is that it iteratively learns its distortion, rotation, and angle individually. The large evolution of self-supervised learning will take place in the world, and so many experts and research apartments work on this topic, and many edge works have been done. For example, self-supervised learning extracts its original and naive manifold features rather than the labeled data intervention. Many methods have shown great accuracy in the computer vision domains, such as Jiasw [21], which is a picture distortion by using contrastive learning. Another self-supervised architecture is called the Moco, and some updated versions are called Moco-v2, also SimLR

[9]. Some large-scale architectures with high performance have also been investigated in the front edge, such as the transformer and the attention mechanism. However, in the industrial aluminum electrolysis domain, self-supervised learning has shown an infant period; a few works have been reported in the literature.

Some studies have been done in the video domains based on self-supervised learning and semi-supervised learning. For example, Xu et al. [22] proposed a self-supervised learning method for the space-time video data and applied it to style transfer. For the time sequence applications, there are also some self-supervised learning methods that emerged [15]. In the literature the unlabeled datasets are learned with self-supervised architectures, and a competitive performance has been achieved. Only data with highly confident labels are combined with original labeled data to train a new model [23]. Developed from [15], a CNN-LSTM model is given for the surgical phase recognition. This method can reduce the burden of manual annotation with the samples [24]. However, in the practical industrial process, it also needs to learn complex knowledge and improved the generation ability and accuracy. Chai et al. [25] improve oblique random forests with dual-incremental learning capacity, providing learning ability, and can update effectively without laborious retraining from scratch. Another customized soft sensor learning is given in [26], which uses a dual attention-based encoder–decoder to learn the sequential information between various input variables and quality output variables. Its efficacy is verified by a real cigarette production benchmark. Chang et al. [27] also proposes a consistent-contrastive network with temporality awareness to obtain robust performance in industrial soft sensor application.

The current proposed state-of-the-art semi-supervised and self-supervised model is based on the self-training paradigm where it first trains a Student Net model on the LSTM-layer representation and uses it as a teacher to generate soft labels (also pseudo-label) on 300M original unlabeled flame hole images in aluminum electrolysis [28,29]. Some recent semi-supervised models have been applied to aluminum electrolysis [30,2]. Compared to semi-supervised learning, self-supervised learning removes the reliability of the labeled data [31]. The similarity of the soft label and permutation are jointly trained with the dual Teacher Net and Student Net [32]. Knowledge distillation as a student model is based on labeled and soft labeled images [33,23]. We iterate this process by putting back the student as the teacher. During the pseudo-label generation, the teacher is not noised, so that the soft labels are as accurate as possible. However, during the student's learning, we inject noise such as dropout, stochastic depth, and data augmentation via a random augmentation to the student to generalize better than the teacher network. The contrastive loss functions are developed to maximize the bound of the soft label and target output.

Based on the above analysis, we summarize the following challenges in the SD identification tasks.

(1) The aluminum electrolysis reduction lacks enough high-quality labels to train the network;
(2) The traditional artificial detection has large objectivity; the large delay, high cost, and inaccuracy exist;
(3) The dynamic relationship of the production variables is time sequential, changing in the whole process;

Lei et al. present a novel self-supervised learning method for aluminum temperature prediction and implement an online framework by the serial LSTM. However, its architecture cannot extract the full features from the time-sequential images dataset. Developed from [34], we propose a novel self-contrastive method to detect the SD in aluminum electrolysis. First, we develop the convolutional neural network in the model to extract the image features as the initial feature representation of the flame hole video, and further, we directly link an LSTM encoder architecture to the top layer. In the top architecture, contrastive learning with a dual teacher net and student net are coordinated to improve the performance at a whole level. To sum up, the following contributions are made in this paper.

(1) A novel deep self-supervised architecture with contrastive learning is proposed with the unlabeled data, an approach for SD identification without requiring tedious manual supervision;
(2) A novel self-supervised loss is constructed, and the knowledge dilation networks are added in the top layers;
(3) Under the labeled data constraint, a robust SD identification method is proposed. The generations and robustness are also evaluated in the paper.

To verify the effectiveness, our method is evaluated by an industrial electrolysis process application; some other comparative methods, which include the supervised methods and semi-supervised methods, are fulfilled in the experiment. Compared to the existing competitive algorithms, the proposed method can achieve state-of-the-art performance.

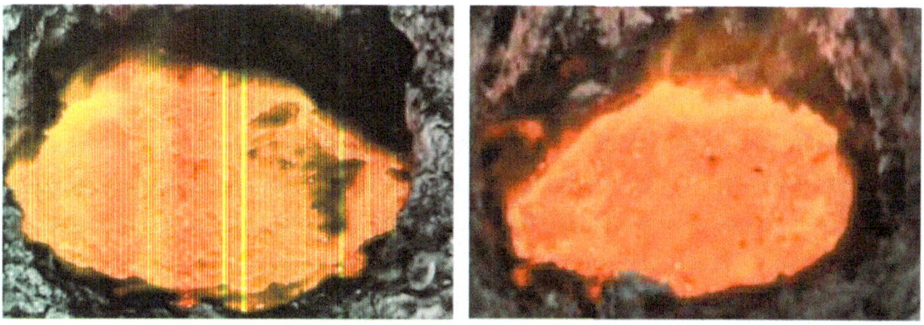

(a) The original flame hole image;

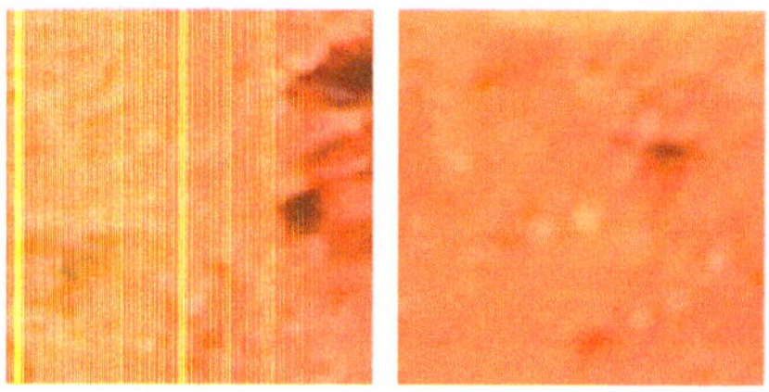

(b) The core deep temperature features after convolution.

Figure 10.1 The deep flame features of unsupervised images after the dilated convolution operation.

10.2 Problem statement

With the difficulty and high cost of acquiring labeled data samples, there is an exponential increase in learning representation in unsupervised scenarios. Self-supervised learning is an active learning method, which has shown great potential to learn the difference between data samples under the constraint of labeled data. For instance, some rotations, angles, and distortions can be learned by the self-supervised model itself. In the proposed CNN-DSSLSTM model, a novel self-supervised loss with a knowledge dilation process is used for the iterative training process. The loss can be described as follows:

$$\mathcal{L}_\theta\left(\boldsymbol{x}_i, \left(\boldsymbol{n}_j, \boldsymbol{n}_k\right)\right) = L\left(\boldsymbol{z}_{i,j}, \boldsymbol{z}_{i,k}\right). \tag{10.1}$$

Mainstream SSL approaches, especially contrastive learning methods, often rely on a series of transformation functions $x_{i,k} = g(x_i, n_k)$ to produce multiple "views" $x_{i,k}$ of input instance x_i. The n_k, $k = 1, \ldots, K$, support an "augmentation nuisances variable" n, representing a group of transformations such as flipping, scaling, cropping, or other augmentations acting on x_i via the function g. The subscript k indexes all possible augmentations, and i indicates the ith instance (each instance corresponds to a specific image input). Without labeling information, SSL relies on the invariance assumption so that the deep feature extractor $z_{i,k} = f(x_i, k)$ is at least insensitive to such a group of augmentation nuisances in the network parameters. Existing work shows that while the deep extractor $f(x_i, k)$ is trained to learn the shared semantics from distorted views x_i, k, such pretrained network parameters effectively benefit further downstream tasks [21]. SSL approaches based on such pairwise instance invariance assumptions can be generally summarized as penalizing some form of inconsistency between paired features z_i, j, z_i, k:

$$\begin{aligned} \mathcal{L}^* &= \mathbb{E}_{\mathbf{x} \in p_{\mathbf{z}}, n \in p^*} \left[\mathcal{L}_\theta(\mathbf{x}, \mathbf{n}) \right] \\ &\approx \frac{1}{N} \sum_{i=1}^{N} \sum_{j,k} L\left(z_{i,j}, z_{i,k}\right) p^* \left(\mathbf{n}_j, \mathbf{n}_k\right). \end{aligned} \tag{10.2}$$

Inspired by [8], the unlabeled images are directly transformed to the target task. The fine-tuned network is then used as a teacher to impute labels to evaluate the soft labels' correctness. Specifically, the following distillation loss with soft labels is

$$L_{\text{distill}} = - \sum_{x_i \in \pi(\cdot)} \left[\sum_{y} P^T\left(y | x_i; \tau\right) \log P^S\left(y | x_i; \tau\right) \right], \tag{10.3}$$

where $P\left(y|x_i\right) = \exp\left(f\left(x_i\right)[y]/\tau\right) / \sum_{y'} \exp\left(f\left(x_i\right)[y']/\tau\right)$, and τ is the scalar parameters. The teacher network that can produce $P^T\left(y|x_i\right)$ in this equation is fixed during the distillation process. The only network to be trained in the student net is to produce $P^S\left(y|x_i\right)$. Assume that we have access to such oracle distribution p, and with equal sampling probability for x_i, the objective function is described by the above equation. Then

$$\mu[f](t, x) = \mu_0(x)\left(1 + \gamma_\mu \rho_p[\rho(t, x)]\right). \tag{10.4}$$

The joint training process of the teacher and student networks can lead to a compact model used to improve the whole performance of the architecture.

10.3 Main framework

SSDLSTM consumes high computation resources for the training process. It motivates us to develop a further framework. Therefore in this section, we introduce a formalization of the proposed CNN-DSSLSTM method, which uses CNN as a feature extractor (see Fig. 10.1).

1) Edge segmentation

To cultivate the burden of the computation, edge segmentation based on a threshold value is used in this architecture. Before the CNN learns the features, the original hole image is transmitted to an edge segmentation unit to extract a core unit. Since tedious annotations of the flame hole images are not available, it is hard to train a multi-scale classifier to separate the categories; instead, we employ self-supervised training. CNN is used to learn some complex feature representations of the original flame images. Assume that the original sequences are denoted as x_{st}. Then a defined updated representation a_t is given as the output of the LSTM layer from the learned CNN-LSTM. Fig. 10.2 describes the self-supervised CNNDLSTM framework for learning the feature representation and its SD application. The building blocks for training are CNN, LSTM, and the final top-layer dual-dilated network. The output of the LSTM, which links from the CNN in a consecutive frame by means of hidden states h_t and the nonlinear activation function, can be rewritten as

$$h_t = \sigma\left(W_h x_t + U_h h_{t-1} + b_h\right), \tag{10.5}$$

and output from the LSTM to the top layer representation can be given:

$$a_t = \sigma\left(W_a h_t + b_h\right). \tag{10.6}$$

In the proposed CNN-SSDLSTM framework, the KL divergence is used to compute the similarity in the training process. Since the label y_i is missing, the hidden features are constructed by the pair (z_i, y_i). Then the log-likelihood function can be written as the following constraint:

$$\begin{aligned} \max \quad & \mathrm{L}^s\left(\boldsymbol{x}_i, \boldsymbol{y}_i\right) = \\ & \mathrm{E}_{q(z)}\left[\ln p\left(\boldsymbol{x}_i, \boldsymbol{y}_i \mid \boldsymbol{z}\right)\right] - KL[q(\boldsymbol{z}) \| p(\boldsymbol{z})], \end{aligned} \tag{10.7}$$

where $\mathrm{E}_q[\cdot]$ is the expectation of the distribution p, and $KL[q(z)\|p(z)] = \int q(z) \ln \frac{q(z)}{p(z)} dz$ is the Kullback–Leibler (KL) divergence to evaluate the dissimilarity between the two labels. Consider the following conditional distribution of x, y, and z:

$$p(z, y \mid x) = p(z \mid y, x) p(y \mid x), \tag{10.8}$$

$$p(\boldsymbol{x} \mid \mathbf{z}, \boldsymbol{y}) = p(\boldsymbol{x} \mid \boldsymbol{z}). \tag{10.9}$$

The static probability is derived as follows:

$$
\begin{aligned}
&\mathrm{L}^{u}\left(\boldsymbol{x}_{i}^{u}\right)= \\
&\mathrm{E}_{p(\boldsymbol{y}|\boldsymbol{x}_{i}^{u})}\left[\mathrm{E}_{p(\boldsymbol{z}|\boldsymbol{x}_{i}^{u},\boldsymbol{y})}\left[\ln p\left(\boldsymbol{x}_{i}^{u}\mid\boldsymbol{z}\right)\right]\right]+\mathrm{H}\left(p\left(\boldsymbol{y}\mid\boldsymbol{x}_{i}^{u}\right)\right) \\
&-\mathrm{E}_{p(\boldsymbol{y}|\boldsymbol{x}_{i}^{u})}\left[\boldsymbol{KL}\left[p\left(\boldsymbol{z}\mid\boldsymbol{y},\boldsymbol{x}_{i}^{u}\right)\|p(\boldsymbol{z})\right]\right] \\
&+\mathrm{E}_{p(\boldsymbol{y}|\boldsymbol{x}_{i}^{u})}\left[\mathrm{E}_{p(\boldsymbol{z}|\boldsymbol{y},\boldsymbol{x}_{i}^{u})}[\ln p(\boldsymbol{y}\mid\boldsymbol{z})]\right] \\
&=\mathrm{E}_{p(\boldsymbol{y}|\boldsymbol{x}_{i}^{u})}\left[\mathrm{L}^{s}\left(\boldsymbol{x}_{i}^{u},\boldsymbol{y}\right)\right]+\mathrm{H}\left(p\left(\boldsymbol{y}\mid\boldsymbol{x}_{i}^{u}\right)\right)].
\end{aligned}
\tag{10.10}
$$

This equation can be expressed as follows:

$$
\begin{aligned}
&\mathrm{KL}\ \left[p\left(\boldsymbol{z},\boldsymbol{y}\mid\boldsymbol{x}_{i}^{u}\right)\|p(\boldsymbol{z},\boldsymbol{y})\right] \\
&=\iint p\left(\boldsymbol{y}\mid\boldsymbol{x}_{i}^{u}\right)p\left(\boldsymbol{z}\mid\boldsymbol{y},\boldsymbol{x}_{i}^{u}\right)\ln\frac{p(\boldsymbol{z}|\boldsymbol{y},\boldsymbol{x}_{i}^{u})p(\boldsymbol{y}|\boldsymbol{x}_{i}^{u})}{p(\boldsymbol{z})p(\boldsymbol{y}|\boldsymbol{z})}d\boldsymbol{z}d\boldsymbol{y} \\
&=\iint p\left(\boldsymbol{y}\mid\boldsymbol{x}_{i}^{u}\right)p\left(\boldsymbol{z}\mid\boldsymbol{y},\boldsymbol{x}_{i}^{u}\right)\ln\frac{p(\boldsymbol{z}|\boldsymbol{y},\boldsymbol{x}_{i}^{u})}{p(\boldsymbol{z})}d\boldsymbol{z}d\boldsymbol{y} \\
&+\iint p\left(\boldsymbol{y}\mid\boldsymbol{x}_{i}^{u}\right)p\left(\boldsymbol{z}\mid\boldsymbol{y},\boldsymbol{x}_{i}^{u}\right)\ln p\left(\boldsymbol{y}\mid\boldsymbol{x}_{i}^{u}\right)d\boldsymbol{z}d\boldsymbol{y} \\
&=\iint p\left(\boldsymbol{y}\mid\boldsymbol{x}_{i}^{u}\right)p\left(\boldsymbol{z}\mid\boldsymbol{y},\boldsymbol{x}_{i}^{u}\right)\ln p(\boldsymbol{y}\mid\boldsymbol{z})d\boldsymbol{z}d\boldsymbol{y} \\
&=\mathrm{E}_{p}\left(\boldsymbol{y}\mid\boldsymbol{x}_{i}^{u}\right)\left[KL\left[p\left(\boldsymbol{z}\mid\boldsymbol{y},\boldsymbol{x}_{i}^{u}\right)\|p(\boldsymbol{z})\right]\right]-\mathrm{H}\left(p\left(\boldsymbol{y}\mid\boldsymbol{x}_{i}^{u}\right)\right) \\
&-\mathrm{E}_{p(\boldsymbol{y}|\boldsymbol{x}_{i}^{u})}\left[\mathrm{E}_{p(\boldsymbol{z}|\boldsymbol{y},\boldsymbol{x}_{i}^{u})}[\ln p(\boldsymbol{y}\mid\boldsymbol{z})]\right],
\end{aligned}
\tag{10.11}
$$

where z denotes the hidden variables, and E_p is the feature expectation of the output variables under the constraint of input x.

2) Contrastive knowledge dilation

In the aluminum electrolysis case study, the performance gap between supervised learning and the scarcity of labeled data should be narrowed. Therefore, to further improve the accuracy of the whole model, the knowledge dilation is applied to the proposed top-layer architecture. The dual teacher and student networks are joined to improve performance and reward the right action.

The output of LSTM can be further tackled with dual contrastive knowledge dilation networks to improve the performance of the proposed model. The specific mathematical equation is

$$
\widehat{y}_t = soft\max\left(W_T a_t + W_S a_t + b_h\right),
\tag{10.12}
$$

where the matrices W_T and W_S are the weight parameters that need to be fine-tuned with the distilled process.

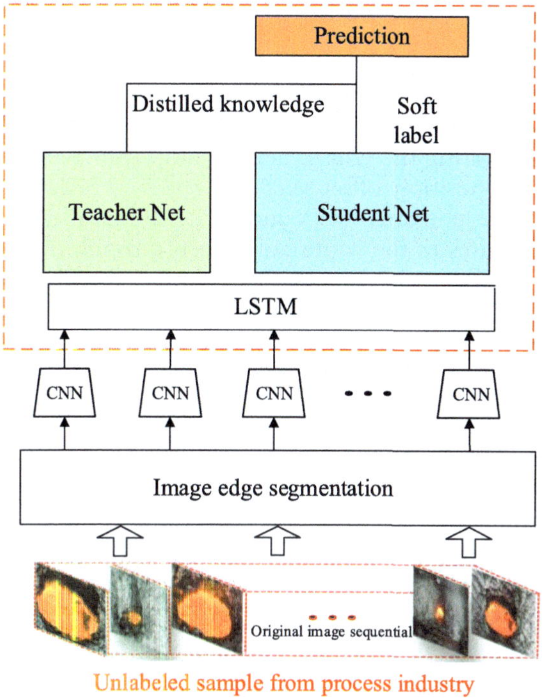

Figure 10.2 The proposed CNN-DSSLSTM framework for industrial electrolysis reduction. The original flame video was divided into pieces of images (1-second frequency sampling, 60 images obtained within 1 min) and then projected to CNN to extract the deep features. Self-supervised LSTM and top-layer dual distillation networks were coordinated to improve the accuracy and performance.

The main flowchart of the proposed algorithm is given in the following steps: 1) Collecting sufficient continuous flame hole videos from the industrial plant scenarios; 2) Using the threshold edge segmentation to divide with the available data ($\sim$1 s per image); 3) Splitting the data with the training, testing, and validation data and initializing the network parameters such as the learning rate α, the batch_size, the hidden layers of LSTM, and the epoch size; 4) Setting up the CNN architecture, the Conv2d layer, and the max-pooling feed word network and then training the CNN weight parameters with the training data; 5) Training the high-level LSTM network based on back-propagation through time (BPTT); 6) Using contrastive learning to train the joint Student Net and Teacher Net to produce soft labels with the corresponding loss given in the above equations; 7) Using the full-concat model to predict the cell temperature in the aluminum electrolysis reduction.

10.4 Numerical examples

The original input is the flame video and position offset vector, and the dimension of flame hole video data is $(n, 1080, 1920, 3)$, where n indicates that the video contains n frames of images, and $(1080, 1920, 3)$ is the image format indicating the width, height, and a number of channels. The dimension of the position offset vector is $(n, 2)$, which means that each frame of the image contains two values. The design idea of the model is to extract the features of the flame hole video through the CNN module, and the position offset vector enters the self-supervised LSTM model after the convolution operation to extract the jitter state features and splicing the output sequence features with the features of the flame hole video after the maximum pooling. Therefore the main reason why the position offset vector first passes through the CNN layer and then enters the self-supervised LSTM is that the position offset vector is not normalized, so the pooling layer is designed to normalize it. Finally, the prediction accuracy of the tank temperature is further improved through the self-distillation model on the top floor. CNN consists of n modules, where n is the number of input frames. The main function of the con display is to extract image features. This proposed CNN-SSDLSTM designs a $v3 - v$ three-layer video structure. For the fully connected layer, the input of the self-supervised LSTM model is 256-dimensional.

Table 10.1 The competitive methods comparison for the temperature identification.

Methods	RMSE	MAE	MAPE	Train_time
LSTM	0.2345	0.2332	0.2621	21.67
RNN	0.3521	0.2546	0.2672	24.89
CNN-LSTM	0.1081	0.1665	0.1339	35.6
CNN-DLSTM	0.0473	0.1823	0.2645	37.99
Ours	**0.0335**	0.0221	**0.0045**	42.86

The dilated learning method is used for improving the accuracy of cell temperature prediction in the application of the plant-wide industry, see Table 10.1. The successive rounds of training improve the temperature accuracy. The highest pot temperature accuracy is up to 93%.

The self-supervised loss is constructed in the network training process. The rotation-augmentation policy is used to augment the amount of unlabeled data. Table 10.2 gives a full comparison of various algorithms. The train_time is evaluated with a second unit, and RMSE, MAE, and MAPE are used to verify the effectiveness of the proposed methods. The commonly competitive methods are RNN, LSTM, CNN-LSTM, CNN-DLSTM, and our methods, and experiments show that the integration

Table 10.2 Training times and variance for the different algorithms with various numbers of unlabeled samples. A: low-temperature zone, B: middle-temperature zone, C: upper-temperature zone. Te_{acc}, Tr_{acc}, and $Tr_{time(s)}$ denote the testing accuracy, training accuracy, and training time, respectively.

Categories	Te_acc			Tr_acc			Tr_time (s)		
	A	B	C	A	B	C	A	B	C
Supervised									
CNN	0.504	0.618	2.626	0.456	0.621	0.481	24.76	24.77	24.77
SimCLR	0.362	1.023	0.128	0.523	0.722	0.813	4.73	5.33	5.22
BYOL	0.233	0.732	0.625	0.616	0.714	**0.881**	6.88	7.31	7.37
Semi-supervised									
SWPPLSR	0.342	0.534	0.636	0.324	0.566	0.698	3.71	3.45	4.23
CNN-LapsELM	0.645	0.423	0.245	0.601	0.711	0.748	3.42	3.44	3.28
Self-supervised									
MoCo	0.674	0.828	0.867	0.086	0.785	0.761	6.21	6.20	6.52
ARTMAP	0.562	0.623	0.648	0.456	0.725	0.737	7.93	8.40	7.22
SSDLSTM	0.693	0.832	0.845	0.656	0.895	0.841	2.59	2.34	2.53
Ours	0.863	**0.934**	**0.905**	0.820	**0.905**	0.891	31.59	25.34	33.53

of self-supervised methods can further lower the RMSE and improve the prediction accuracy at the same time. The accuracy of the final proposed model validation set is 0.823, and the accuracy of the position offset CNN-SSDLSTM model validation set is 0.845, respectively. In addition, performances of CNN-LSTM and CNN-DLSTMs outperform the basic RNN and LSTM units, which shows that CNN can learn better information from the flame hole videos. Our CNN-SSDLSTM achieves the lowest RMSE and MAE, 0.0335 and 0.0221, respectively, which is the result of CNN, self-supervised loss, and constructive design, although it also costs a longer training time due to the independent CNN training process, see Fig. 10.3.

In the comparison experiment, after two iteration rounds, rapid convergence and increase of accuracy are observed. The contrastive joint training of the top dual network learns a superior representation of the unlabeled flame images. Compared to the traditional RNN and LSTM unit, the prediction performance has increased by a large margin, up to 0.92.

In the above experiments, the proposed CNN-SSDLSTM provides state-of-the-art performance compared with the other competitive methods. The advantages of the proposed method can be specified as follows:

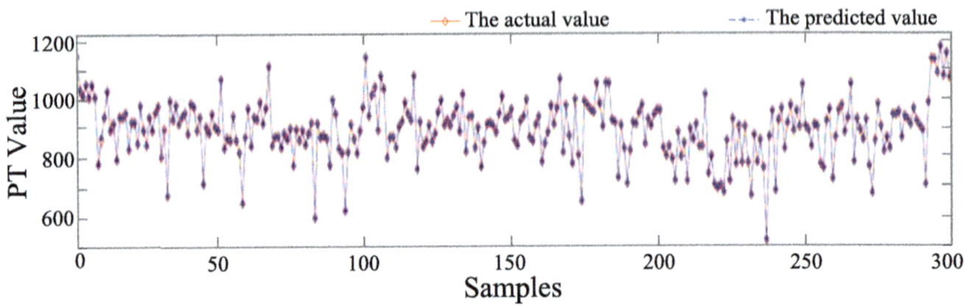

Figure 10.3 CNN-SSDLSTM testing error comparison with various numbers of unlabeled samples for different algorithms. Other comparisons are provided in Chapter 9.

- Economic profit saving. The proposed method is a fully data-driven method, which uses the data to train the algorithms rather than the physical devices with a lot of profit savings.
- Online real-time prediction. The method proposes an online real-time prediction for SD in the aluminum reduction cell. It can be also extended to other similar industrial plant-wide applications.
- Breaking the constraint of labeled samples. This method uses self-supervised contrastive learning to narrow the gap between supervised learning and the scarcity of labeled samples. Since practical industrial plants are mainly the categories of unlabeled data, the generation of the proposed method is greatly improved.

10.5 Conclusions

In this framework, an improved self-supervised LSTM method is proposed for aluminum temperature prediction. The deep architecture of the proposed algorithm leverages the loss function with contrastive learning. Further, a novel unit with a self-supervised loss function is used. The knowledge dilation with dual teacher net and student net is also jointly trained for the proposed architecture. The experimental results demonstrate that the performance outperforms the other existing comparisons. The proposed self-supervising function is vividly used in the CNN-SSDLSTM models and has received competitive experimental and application verification. Some extended exploration of this method can be in fault soft sensor domains, which have the great potential to deal with fault prediction under the limitation of sufficient labels.

References

[1] Ziqing Deng, Xiaofang Chen, Shiwen Xie, Yongfang Xie, Yubo Sun, Distributed process monitoring based on joint mutual information and projective dictionary pair learning, Journal of Process Control 106 (2021) 130–141.

[2] Ziqing Deng, Xiaofang Chen, Shiwen Xie, Yongfang Xie, Hongliang Zhang, Semi-supervised discriminative projective dictionary pair learning and its application for industrial process monitoring, IEEE Transactions on Industrial Informatics (2022).

[3] Keke Huang, Yiming Wu, Cheng Long, Hongquan Ji, Bei Sun, Xiaofang Chen, Chunhua Yang, Adaptive process monitoring via online dictionary learning and its industrial application, ISA Transactions 114 (2021) 399–412.

[4] Zhaoke Huang, Chunhua Yang, Xiaofang Chen, Keke Huang, Yongfang Xie, Adaptive over-sampling method for classification with application to imbalanced datasets in aluminum electrolysis, Neural Computing & Applications 32 (11) (2020) 7183–7199.

[5] Zhaoke Huang, Chunhua Yang, Xiaofang Chen, Xiaojun Zhou, Guo Chen, Tingwen Huang, Weihua Gui, Functional deep echo state network improved by a bi-level optimization approach for multivariate time series classification, Applied Soft Computing 106 (2021) 107314.

[6] Weichao Yue, Weihua Gui, Xiaofang Chen, Zhaohui Zeng, Yongfang Xie, Evaluation strategy and mass balance for making decision about the amount of aluminum fluoride addition based on superheat degree, Journal of Industrial and Management Optimization 16 (2) (2020) 601.

[7] Filippo Aleotti, Fabio Tosi, Li Zhang, Matteo Poggi, Stefano Mattoccia, Reversing the cycle: self-supervised deep stereo through enhanced monocular distillation, in: European Conference on Computer Vision, Springer, 2020, pp. 614–632.

[8] Ting Chen, Simon Kornblith, Kevin Swersky, Mohammad Norouzi, Geoffrey E. Hinton, Big self-supervised models are strong semi-supervised learners, Advances in Neural Information Processing Systems 33 (2020) 22243–22255.

[9] Xinlei Chen, Haoqi Fan, Ross Girshick, Kaiming He, Improved baselines with momentum contrastive learning, arXiv preprint, arXiv:2003.04297, 2020.

[10] Alexander Kolesnikov, Xiaohua Zhai, Lucas Beyer, Revisiting self-supervised visual representation learning, in: Proceedings of the IEEE/CVF Conference on Computer Vision and Pattern Recognition, 2019, pp. 1920–1929.

[11] Yongxiang Lei, Hamid Reza Karimi, Lihui Cen, Xiaofang Chen, Yongfang Xie, Processes soft modeling based on stacked autoencoders and wavelet extreme learning machine for aluminum plant-wide application, Control Engineering Practice 108 (2021) 104706.

[12] Yongxiang Lei, Xiaofang Chen, Yongfang Xie, An improved cell situation identification approach with convolutional neural network and wavelet extreme learning machine, Proceedings of the Institution of Mechanical Engineers. Part I, Journal of Systems and Control Engineering (2020) 0959651820935667.

[13] Longlong Jing, Yingli Tian, Self-supervised visual feature learning with deep neural networks: a survey, IEEE Transactions on Pattern Analysis and Machine Intelligence (2020).

[14] Jingkuan Song, Hanwang Zhang, Xiangpeng Li, Lianli Gao, Meng Wang, Richang Hong, Self-supervised video hashing with hierarchical binary auto-encoder, IEEE Transactions on Image Processing 27 (7) (2018) 3210–3221.

[15] Gaurav Yengera, Didier Mutter, Jacques Marescaux, Nicolas Padoy, Less is more: surgical phase recognition with less annotations through self-supervised pre-training of CNN-LSTM networks, arXiv preprint, arXiv:1805.08569, 2018.

[16] Xiaohua Zhai, Avital Oliver, Alexander Kolesnikov, Lucas Beyer, S4L: self-supervised semi-supervised learning, in: Proceedings of the IEEE/CVF International Conference on Computer Vision, 2019, pp. 1476–1485.

[17] Mariana-Iuliana Georgescu, Antonio Barbalau, Radu Tudor Ionescu, Fahad Shahbaz Khan, Marius Popescu, Mubarak Shah, Anomaly detection in video via self-supervised and multi-task learning, in: Proceedings of the IEEE/CVF Conference on Computer Vision and Pattern Recognition, 2021, pp. 12742–12752.
[18] Jean-Bastien Grill, Florian Strub, Florent Altché, Corentin Tallec, Pierre Richemond, Elena Buchatskaya, Carl Doersch, Bernardo Avila Pires, Zhaohan Guo, Mohammad Gheshlaghi Azar, et al., Bootstrap your own latent: a new approach to self-supervised learning, Advances in Neural Information Processing Systems 33 (2020) 21271–21284.
[19] Pierre Sermanet, Corey Lynch, Yevgen Chebotar, Jasmine Hsu, Eric Jang, Stefan Schaal, Sergey Levine, Google Brain, Time-contrastive networks: self-supervised learning from video, in: 2018 IEEE International Conference on Robotics and Automation (ICRA), IEEE, 2018, pp. 1134–1141.
[20] Tian Shi, Liuqing Li, Ping Wang, Chandan K. Reddy, A simple and effective self-supervised contrastive learning framework for aspect detection, arXiv preprint, arXiv:2009.09107, 2020.
[21] Ishan Misra, Laurens van der Maaten, Self-supervised learning of pretext-invariant representations, in: Proceedings of the IEEE/CVF Conference on Computer Vision and Pattern Recognition, 2020, pp. 6707–6717.
[22] Kai Xu, Longyin Wen, Guorong Li, Honggang Qi, Liefeng Bo, Qingming Huang, Learning self-supervised space-time CNN for fast video style transfer, IEEE Transactions on Image Processing 30 (2021) 2501–2512.
[23] Chuanguang Yang, Zhulin An, Linhang Cai, Yongjun Xu, Hierarchical self-supervised augmented knowledge distillation, arXiv preprint, arXiv:2107.13715, 2021.
[24] Taewook Kim, Ha Young Kim, Forecasting stock prices with a feature fusion LSTM-CNN model using different representations of the same data, PLoS ONE 14 (2) (2019) e0212320.
[25] Zheng Chai, Chunhui Zhao, Multiclass oblique random forests with dual-incremental learning capacity, IEEE Transactions on Neural Networks and Learning Systems 31 (12) (2020) 5192–5203.
[26] Liangjun Feng, Chunhui Zhao, Youxian Sun, Dual attention-based encoder–decoder: a customized sequence-to-sequence learning for soft sensor development, IEEE Transactions on Neural Networks and Learning Systems 32 (8) (2020) 3306–3317.
[27] Shuchao Chang, Chunhui Zhao, Ke Li, Consistent-contrastive network with temporality-awareness for robust-to-anomaly industrial soft sensor, IEEE Transactions on Instrumentation and Measurement 71 (2021) 1–12.
[28] Liangjun Feng, Chunhui Zhao, Biao Huang, Adversarial smoothing tri-regression for robust semi-supervised industrial soft sensor, Journal of Process Control 108 (2021) 86–97.
[29] Yalin Wang, Haibing Yang, Xiaofeng Yuan, Yuri A.W. Shardt, Chunhua Yang, Weihua Gui, Deep learning for fault-relevant feature extraction and fault classification with stacked supervised auto-encoder, Journal of Process Control 92 (2020) 79–89.
[30] Jie Wang, Yongfang Xie, Shiwen Xie, Xiaofang Chen, Optimization of aluminum fluoride addition in aluminum electrolysis process based on pruned sparse fuzzy neural network, ISA Transactions (2022).
[31] Le Zhou, Junghui Chen, Zhihuan Song, Zhiqiang Ge, Semi-supervised PLVR models for process monitoring with unequal sample sizes of process variables and quality variables, Journal of Process Control 26 (2015) 1–16.
[32] Jun Yue, Leyuan Fang, Hossein Rahmani, Pedram Ghamisi, Self-supervised learning with adaptive distillation for hyperspectral image classification, IEEE Transactions on Geoscience and Remote Sensing 60 (2021) 1–13.
[33] Guodong Xu, Ziwei Liu, Xiaoxiao Li, Chen Change Loy, Knowledge distillation meets self-supervision, in: European Conference on Computer Vision, Springer, 2020, pp. 588–604.

[34] Yongxiang Lei, Hamid Reza Karimi, Xiaofang Chen, A novel self-supervised deep LSTM network for industrial temperature prediction in aluminum processes application, Neurocomputing 502 (2022) 177–185.

CHAPTER 11

A novel probabilistic prediction framework based on Bayesian machine learning: BLSTM

11.1 Introduction

The aluminum electrolysis industry stands as a pillar of the national economy, serving as a crucial raw material for numerous fields, including astronomy, chemistry, and medicine [1–4]. This process involves complex reactions, focusing primarily on the production of pure aluminum [5–7]. The electrolysis cell is the main reaction unit for the whole process, which is usually a harsh environment coupled with multiple variables with high temperature and pressure and strong magnification [8–10]. One key quality variable crucial to the primary reaction is the cell temperature. Maintaining it within a dynamic range is of utmost importance in practical applications as deviations can lead to adverse effects such as electrolysis production cell erosion, shortened bath lifespan, or even safety hazards [11]. In contrast, lower temperatures can interfere with the aluminum reaction, resulting in a decreased production efficiency [12]. Accurate temperature prediction has immense significance in real-world scenarios. It provides essential information about the overall cell situation, enabling better control strategies, production monitoring, and proactive avoidance of critical faults.

In terms of aluminum production, temperature detection is currently carried out by workers using traditional physical devices, such as thermocouples or high-temperature infrared pyrometers. However, these devices have significant drawbacks. Firstly, their detection accuracy is limited by the low sampling frequencies and the quality of the original materials used in the devices. Alternatively, high-accuracy devices may incur substantial costs [13,14]. Furthermore, this temperature detection approach is conducted offline, providing only local temperature readings. As a result, it fails to capture the dynamic complexity of the entire cell temperature field, which exhibits different temperature distributions at various probe locations. With the advent of databases and machine learning, more cost-effective prediction methods have emerged for temperature

Learning-Based Soft Sensing and Predictions for Process Industries
https://doi.org/10.1016/B978-0-44-336759-5.00020-2

tasks. For example, Huang et al. [2] proposed an improved TOPSIS-based multi-criteria decision-making approach to evaluate the working condition of aluminum reduction cells. However, their investigation did not evaluate the cell temperature index during the training process. Similarly, a time series clustering method with extra validation was recently proposed, which could identify an unknown cell condition using a newly developed validity index. Nonetheless, their method was focused on the clustering task and did not facilitate real-time temperature forecasting [15]. To address the aforementioned limitation, a functional deep echo state network with a bi-level optimization approach is proposed for aluminum electrolysis multivariate classification [16]. By incorporating the state-transition algorithm the model performance is improved. However, their primary emphasis was on the classification task, largely overlooking the regression of cell temperature.

Over the last decade, several data-driven methods have been proposed for aluminum electrolysis applications. For instance, Lei et al. [17] developed a semi-supervised extreme learning machine that leveraged a fusion strategy of data features and image features for superheat identification. Although their model exhibited enhanced feature representation compared to traditional methods, it required high-quality data for training, and the training cost was also significant. Building upon this foundation, further developments have been reported in [18,19]. These methods were based on historical data linkage and aluminum electrolysis reaction mechanism, leading to the creation of data-driven soft sensor techniques that provided valuable insights for the process industry. Recently, a self-supervised temporal temperature prediction method was reported, employing cutting-edge self-supervised learning methods to reduce the reliance on labeled data [20]. Their framework used long short-term memory as the core unit and introduced a new self-supervised loss for the training process. This approach achieved competitive performance despite the constraints of limited samples. Another breakthrough in the aluminum electrolysis temperature prediction was reported in [21]. The method involved constructing a dilated contrastive learning process in the network architecture design, effectively reducing the training cost. Additionally, the introduction of unlabeled flame images dataset was a notable contribution, mitigating the time-consuming nature of training large-scale unsupervised images [22]. The dilated process not only reduced training costs but also yielded satisfactory temperature prediction performance. However, it is worth noting that both of the aforementioned works were based on deterministic prediction, which may lack robustness and flexibility during model deployment. Moreover, they overlooked the influence of noises and original data distribution during training, potentially limiting the model generation ability and leading to prediction oscillation when encountering an outlier.

Some earlier works on industrial temperature predictions were predominantly relied on deterministic methods [23–25]. However, these conventional prediction techniques, such as nonlinear auto-regressive models, typically offer only short-term predictions and overlook uncertainty information [26]. Pan et al. [27,28] proposed a new temperature compensation method for ironmaking dust measurement, but it required the installation and functionality of an external high-temperature thermal monitor. Although compensation-based temperature measurement methods have been implemented [27], the need for temperature detection is crucial for various industrial processes. Studies focused on aluminum electrolysis prediction and maintenance, based on the material application mechanism, have achieved practical implementation with satisfactory performance [29,30]. Another mainstream to temperature monitoring method is the use of soft sensor. For example, in the rotary kiln calcination application, a multi-model fusion method has been employed [31]. Additionally, soft sensors have been applied for molten steel temperature [32] and extrapolation of iron-making blast furnace temperature [33]. More recently, discussions about soft sensors for industrial process prediction have emerged in the literature [34–36]. These studies investigated the superior performance of long short-term memory (LSTM) in predicting key quality variables for various industrial processes.

Over the past decade, the emergence and development of machine learning techniques have led to the widespread adoption of data-driven models in the literature, offering promising avenues for uncertainty temperature prediction [37,38]. In one such study, [39], an LSTM network was proposed for temperature forecasting and compared to the nonlinear auto-regressive neural network (NARNN) and group method of data handling network [40]. Their method improved the prediction performance; however, it did not consider the robustness of predictions. In recent times, probabilistic Bayesian neural networks have garnered significant attention [41,42]. This process introduces Gaussian noise to the original distribution, resulting in variant feature information. This technique has found wide application in the vision and text tasks [43,44]. However, despite these advancements, the consideration of uncertainty remains based on deterministic methods, lacking the redundancy of uncertainty. As a result, the prediction method could yield an infinite pulse when subjected to abnormal disturbances during testing sampling, potentially leading to control device malfunctions and increased production losses. Bayesian layers seek to introduce uncertainty on its weights by sampling them from a distribution parameterized by trainable variables on each feedforward operation [45]. This allows us not just to optimize the performance metrics of the model, but also gather the uncertainty of the network predictions over a specific datapoint (by sampling it much times and measuring the dispersion) and aimingly reduce as much as possible the variance of the network over the prediction, making it possible to know how much of uncertainty we still have over the label if we try to model it in function of our specific datapoint [46].

This chapter aims to overcome the limitations of previous deterministic methods in industrial aluminum electrolysis temperature prediction by developing a novel uncertainty prediction approach. Unlike conventional approaches, this study explores the implementation of a probabilistic sequence model to quantify both epistemic and aleatoric uncertainties during the temperature forecasting process. Particularly, the investigation focuses on using the probabilistic sequence neural network, which is capable of non-casual temperature prediction, a novel approach in this context. The improved probabilistic neural network offers more accurate predictions while accounting for uncertainty. Additionally, this work goes beyond previous studies, which primarily focused on linear-projection temperature prediction, by incorporating a nonlinear real-time uncertainty strategy. The proposed method harnesses the capabilities of the machine learning models to tackle complex sequence prediction while considering various forms of uncertainties within a Bayesian framework, a novel aspect in this field, to the best knowledge of the authors. In summary, the main bullets of this paper are as follows.

- Compared with traditional conservative temperature prediction methods [47,20], a novel prediction model is proposed. This model processes sequence information using an LSTM network and subsequently uses a probabilistic Bayesian neural network to learn different distributions.
- District from previous mainstream deterministic prediction methods [36,34], the presented model implements a dynamic robust prediction, providing valuable reference information for later control processes. As a result, the full-scale temperature prediction avoids over-conservative estimation, offering more accurate results.

- The method's foundation lies in the probabilistic approach, a cutting-edge technique that achieves uncertainty quantification. Building upon this accurate prediction, the proposed strategy outperforms existing deterministic sequence prediction competitors, such as artificial neural network (ANN) and long short-term memory (LSTM).
- To the best of the authors' knowledge, the presented work lays the groundwork for uncertain prediction criteria within the aluminum industry, which also can be extended for other processes industry. The practical performance comparison with existing methods is evaluated in the aluminum reduction cell, and experimental results show that the error ratio is bounded within 5‰ with two typical cell benchmarks.

11.2 Problem statement

In aluminum electrolysis reduction production, the cell temperature, also known as the electrolyte temperature, plays an important role as a key indicator for monitoring the process and facilitating superheat identification. Accurate temperature prediction is of utmost importance, as it empowers us to take proactive actions or implement effective control strategy. By doing so we can prevent the aluminum electrolysis cell from entering an abnormal situation and minimize unnecessary economic losses. The temperature is linked with the other process variable, which forms the foundation for developing a data-driven temperature prediction algorithm.

Another core element in the aluminum electrolysis plant is that the cell temperature is a dynamic sequence process. Specifically, the temperature is affected by the historical temperature observation and other key quality process variables factor. The most straightforward way to implement the prediction is based on the auto-regressive model [48]. Moreover, the complexity of the temperature time-sequence series extends beyond a mere relationship with its own univariate historical temperature. It is also coupled with other process variables, such as the initial crystal temperature. The consideration of reaction mechanism analysis draws us to the point that the temperature is also influenced by the complex thermal equilibrium, which is difficult to compute directly. The next stage involves selecting the most relevant variables that are strongly correlated with the temperature. All the process variables, along with their descriptions, are summarized in Table 3.1. The correlation analysis is conducted before formulating our problem, and the results of this analysis are presented.

Therefore, under the constraints of these key quality variables, the cell temperature prediction problem is finally formulated as follows.
Given: historical temperature series

$$\mathbf{x_{t-1}} = < T(t-m-1), T(t-m), \ldots, T(t-1) >$$

and the other key quality variables, where t is the historical time window instant, m is the total historical time interval for the subsequent prediction of the temperature, and l is the number of selected key quality variables. Then the final augmented input for the temperature forecasting is

$$\mathbf{x_{aug}}(t-1) = \begin{bmatrix} T(t-m-1), T(t-m), \ldots, T(t-1) \\ u_1(t-m-1), u_1(t-m), \ldots, u_1(t-1) \\ \cdots\cdots \\ u_l(t-m-1), u_l(t-m), \ldots, u_l(t-1) \end{bmatrix}.$$

Predict: The future time sequence temperature

$$\mathbf{y_t} = < T(t), T(t+1), \ldots, T(t+N_p) >,$$

where $N_p = T - t + 1$ is the preceding horizon served as the temperature target.

11.3 Main framework

The accurate temperature prediction is critical in aluminum production. However, traditional model ignores the uncertainty. We propose a novel probabilistic framework BLSTM, given in Fig. 11.1. Let x^0 be a multivariate training vector drawn from an input space $\mathcal{X} = \mathbb{R}^D$ with probability density function (PDF) of $p_\theta(\mathbf{x}^0)$. Probabilistic models are a type of latent variable model represented by the PDF $p_\theta(\mathbf{x}^0)$, where latent variables $\mathbf{x}^1, \ldots, \mathbf{x}^N$ of dimension $\mathbb{R}^D$ are integrated over $p_\theta(\mathbf{x}^0) := \int p_\theta(\mathbf{x}^{0:N})d\mathbf{x}^{1:N}$. The approximate posterior $q(\mathbf{x}^{1:N} \mid \mathbf{x}^0)$ is a variational representation:

$$q(\mathbf{x}^{1:N} \mid \mathbf{x}^0) = \Pi_{n=1}^{N} q(\mathbf{x}^n \mid \mathbf{x}^{n-1}). \tag{11.1}$$

This equation is fixed to a Markov chain (called the forward process). Considering the covariate, the equation can be expressed in a more precisely way:

$$q(\mathbf{x}^{1:N} \mid \mathbf{x}^0, \mathbf{z}^0) = \Pi_{n=1}^{N} q(\mathbf{x}^n \mid \mathbf{x}^{n-1}, \mathbf{z}^{n-1}) = \Pi_{t=1}^{T} q(\mathbf{x}_t \mid \mathbf{h}_t), \tag{11.2}$$

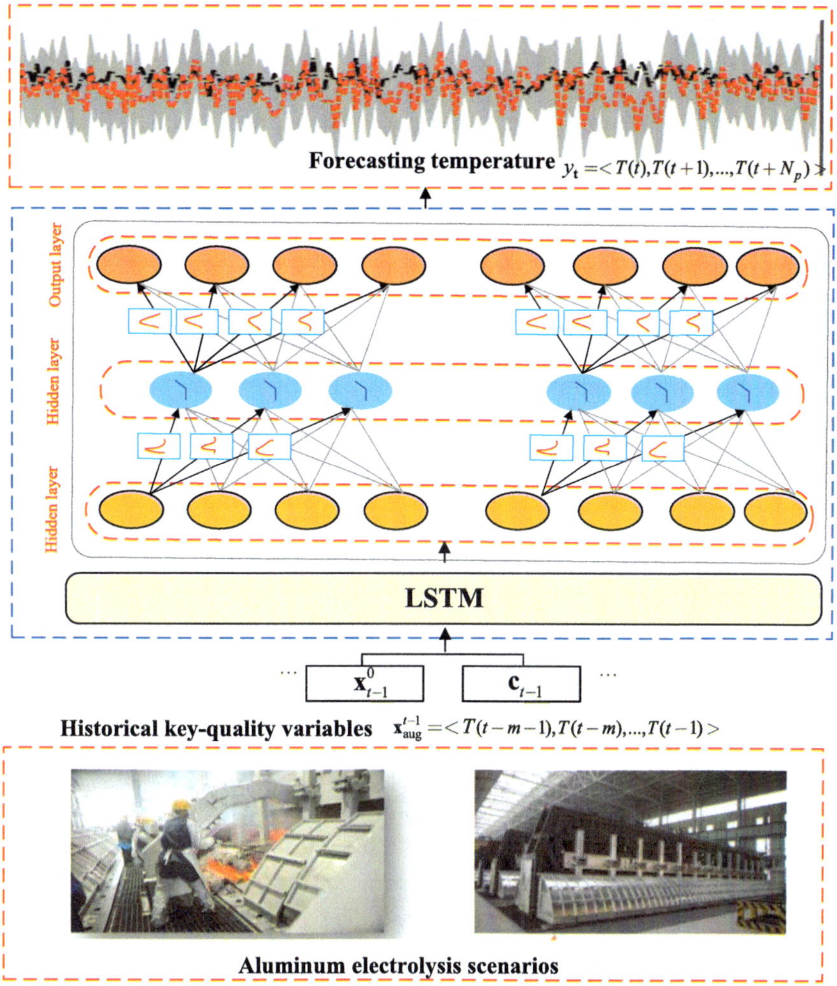

Figure 11.1 The framework of the proposed ML-based wave excitation moment prediction. An ML-based temporal conditioned probabilistic model at some time $t-1$ depicting the fixed forward process.

where $\mathbf{h_t}$ is the output of LSTM, which is also a nonlinear LSTM cell f function with parameter θ. Eq. (11.2) is computed as follows:

$$\mathbf{h_t} = f(\mathbf{h_{t-1}}, \mathbf{x_{t-1}}, \mathbf{z_t}, \theta). \tag{11.3}$$

For the temperature case study, the conditional distribution is represented as

$$q_{\mathcal{X}}(\mathbf{x}^0_{t_0:T} \mid \mathbf{x}^0_{1:t_0-1}, \mathbf{c}_{1:T}) = \Pi^T_{t=t_0} q_{\mathcal{X}}(\mathbf{x}^0_t \mid \mathbf{x}^0_{1:t-1}, \mathbf{c}_{1:T}). \tag{11.4}$$

Because the covariates are known for all time points, each factor is learned through the conditional denoising model introduced earlier. Parameter training is carried out by randomly sampling cell temperature sequences and combining with the prediction steps from the training data, and then optimizing the parameters θ that minimize the model negative log-likelihood

$$L(\theta) = \sum_{t=t_0}^{T} - \log p_\theta(\mathbf{x}_t^0 \mid \mathbf{h}_{t-1}), \tag{11.5}$$

where the hidden state $\mathbf{h}_{t_0-1}$ is obtained by running the LSTM on the chosen time-series temperature window. Therefore we can obtain the predicted temperature through ancestral sampling as follows:

$$\hat{\mathbf{x}}_t \sim \Pi_{t=t_0}^{T} q_X(\cdot \mid \hat{\mathbf{h}}_{t-1}), \tag{11.6}$$

where $\hat{\mathbf{x}}_t = \mathbf{x}_t$ and $\hat{\mathbf{h}}_{t-1} = \mathbf{h}_{t-1}$ for the initial samples. The predictive range is recursively implemented by N steps.

11.4 Numerical examples

In this section, we use the proposed algorithm to predict the pot temperature of aluminum electrolysis. The other machine learning prediction methods are used as competitors. Two typical aluminum electrolysis cells labeled with reactors #8888 and #8889 are used to test the method. Firstly, we introduce the experimental setting, providing the details of the setup and parameters used in the study. Following this, we extensively discuss and analyze the prediction performance, presenting comprehensive insights into the results obtained from our method. First, the proposed prediction method with uncertainty qualification is carried out on the whole time domain. The results are shown in Fig. 11.2(a) and (b) for two typical aluminum reduction cells, which are denoted as reactors #8888 and #8889, respectively. Fig. 11.2 presents the temperature predicted by the proposed Bayesian machine learning, deterministic neural network, and traditional long short-term memory, including both the mean values and the 95% confidence interval. As shown, the temperature prediction in the testing samples (red dotted line) matches very well with the reference values (marked in black dashed-dotted line), and the predicted 95% confidence interval (gray zone) characterizes the prediction uncertainty accurately throughout the prediction time domain. In particular, the prediction zone

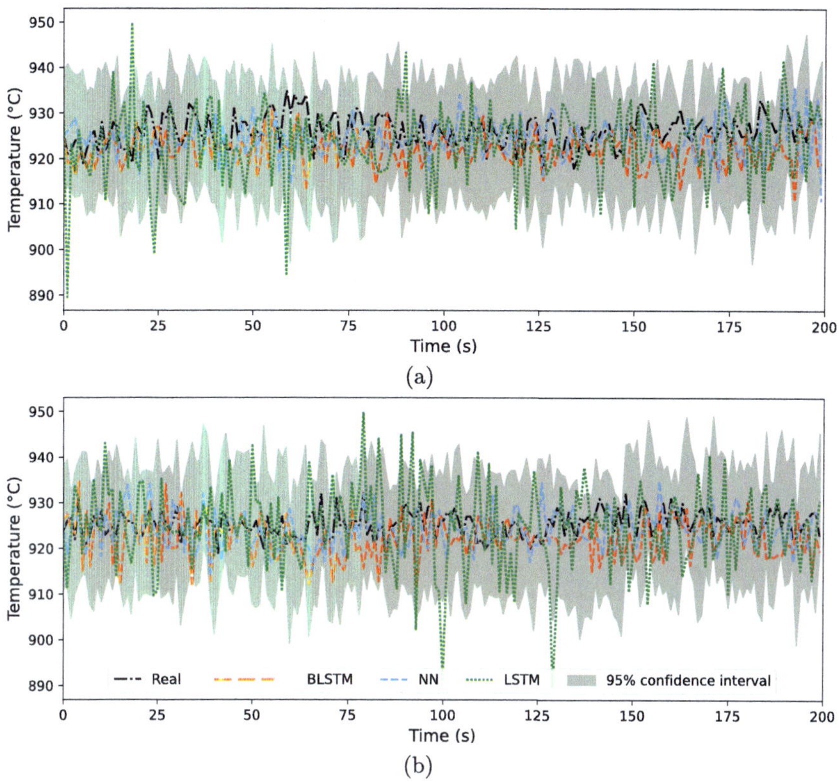

Figure 11.2 Various aluminum reactor comparisons with various cell temperature prediction algorithms. Particularly, to make competitive comparisons, all the results are based on the testing datasets with same hyper-parameters setting (a) Reactor #8889, (b) Reactor #8888.

with 95% confidence degree covers the predicted mean and reference values, which means that the temperature prediction falls within the predicted 95% confidence interval.

For comparison purposes, the deterministic methods by the NN-based and traditional LSTM are both examined. As shown in Fig. 11.2, the proposed method matches the real temperature better that the two deterministic comparisons. The presented method shows a great performance compared to the deterministic approach, while it is only slightly different from the LSTM prediction, which is reasonable as the probabilistic LSTM is validated as a powerful learner for the time sequence tasks. Thus the unique benefit of the method proposed in this work is that it can predict the prediction uncertainty, as shown in Fig. 11.3(a) and

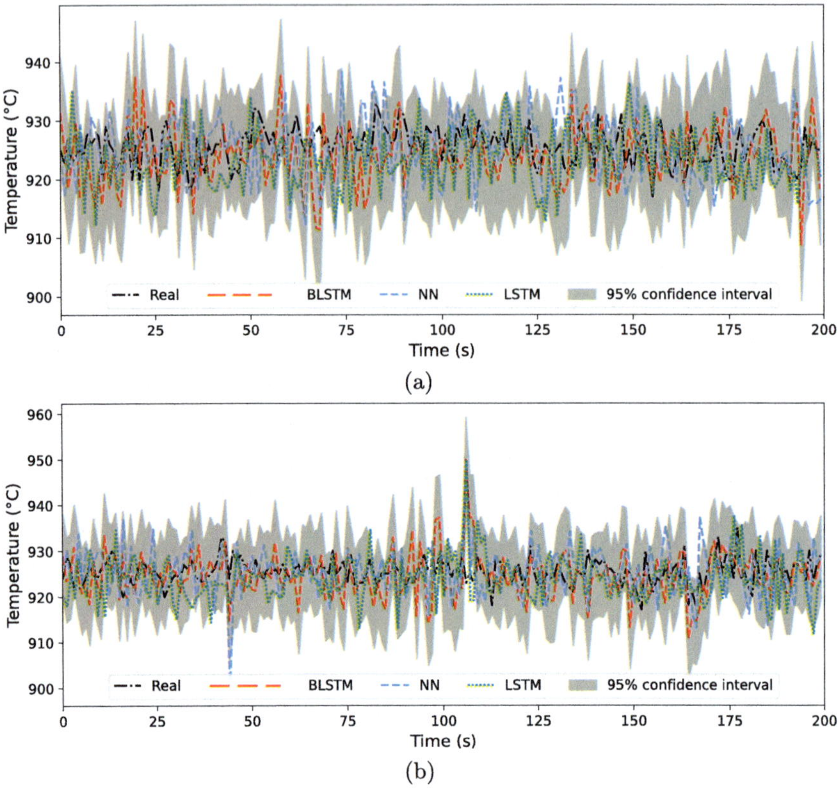

Figure 11.3 Different aluminum reactor comparisons with the different cell temperature prediction algorithms, particularly, to make the competitive comparisons, all the results are based on the training datasets with same hyper-parameters setting; (a) Reactor #8889 (b) Reactor #8888.

(b), along with the prediction averaged value. In this way, epistemic and aleatoric uncertainties are both obtained. The 95% confidence interval provides valuable information for controlling the aluminum reduction cell, ensuring robust performance. In contrast, traditional temperature predictions can only offer point estimation, lacking the comprehensive uncertainty assessment that the confidence interval provides.

Moreover, an obvious advantage of the proposed method is the oscillation suppression throughout the prediction horizon. As intuitively shown in Fig. 11.4(a), the NN-based method and LSTM tend to overpredict the temperature and cause greater root mean square error and oscillation, whereas the prediction from the proposed probabilistic Bayesian

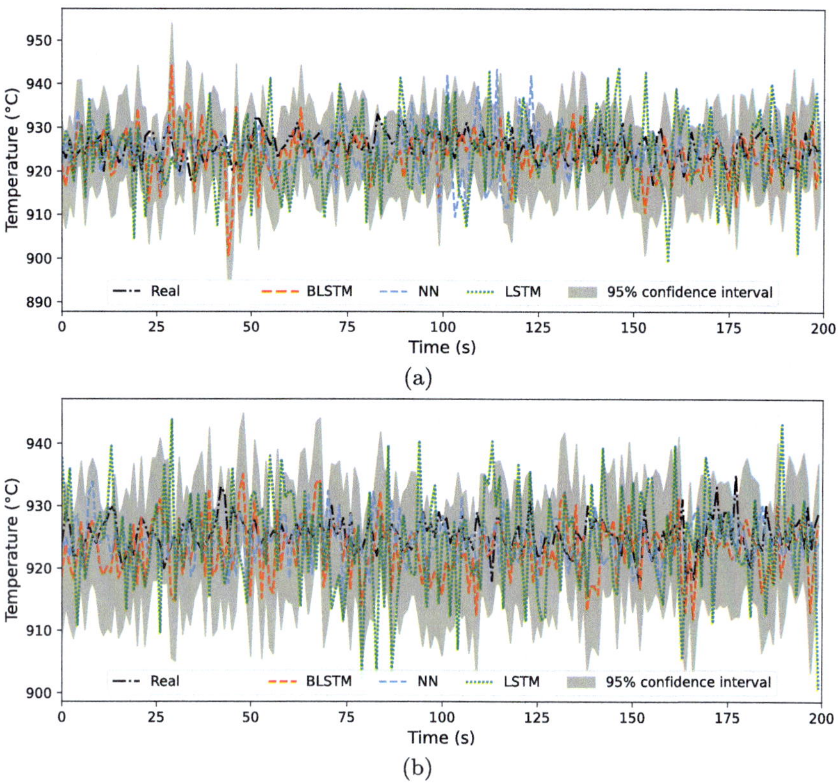

Figure 11.4 Various aluminum reactor comparisons with various cell temperature prediction algorithms. Particularly, to make the competitive comparisons, all the results are based on the validation datasets with same hyper-parameters setting (a) Reactor #8889, (b) Reactor #8888.

LSTM tends to predict the temperature in a more conservative way. This result should be explained by the fact that the proposed methods should be stronger learners with high interpretation ability to get the information about these historical key quality observations, thus reducing the error bound of the confidence interval. For reactor #8889 shown in Fig. 11.4(b), similar charismatics are shown, while achieving a better 95% confidence interval, because the gray error defined by the upper and lower boundaries is smaller than in the case of reactor #8888. Therefor, a reliable temperature prediction with uncertainty quantification is implemented, see Table 11.1. In addition to the cell temperature prediction with quantified uncertainty, the determination of the accuracy indexes is also achieved by the proposed method. To make the full ablation and evaluate the overall prediction

Table 11.1 Network architecture and hyper-parameters for the comparison experiments.

Method	lr	scale factor	β	epoches
NN	0.001	0.9	–	2×10^3
LSTM	0.001	0.9	–	2×10^3
BLSTM	0.001	0.9	0.9	2×10^3

error, firstly, the mean square and root square errors for the dynamic zone's prediction and reference temperature values are calculated, which are the direct reflection of the cell temperature prediction performance and effectiveness. The mean square and root mean square errors for the aluminum electrolysis temperature prediction are defined as

$$MSE = \frac{1}{T_{\max} - T_a} \sum_{T_a \leqslant i \leqslant T_{\max}} (y_i^* - y_i)^2, \tag{11.7}$$

$$RMSE = \sqrt{\frac{1}{T_{\max} - T_a} \sum_{T_a \leqslant i \leqslant T_{\max}} (y_i^* - y_i)^2}, \tag{11.8}$$

where $[T_{\max}, T_a]$ is the prediction range by NN-based methods and deterministic LSTM benchmark, y_i^* and y_i are the predicted and reference values of the cell temperature at the ith time instant. Following Fig. 11.4, the corresponding intuitive RMSE results are shown in Fig. 11.5. The dynamic RMSEs of the proposed Bayesian LSTM, deterministic NN-based method, and LSTM are included. It is worth pointing out that the performance of the proposed uncertain LSTM is superior in the different cells, with the lowest RMSE obtained. More specifically, in the case of Reactor #8888 (Fig. 11.6(a)), the NN-based deterministic prediction is usually with high oscillation in many instants, which finally increases the averaged root mean square error for the forecasting horizon and degrades the method performance. Compared to the NN-based approach, the LSTM achieves a better prediction and a corresponding lower rmse, which provides the information that the time sequence is more approximate for the industrial aluminum temperature application. The proposed Bayesian LSTM is marked in red, and as shown, the RMSE is wholly bounded with correspondence to the prediction horizon, and the lowest RMSE is presented at the same time. For reactor #8889, as another comparison, the results are shown in Fig. 11.6(b). The whole error distribution of the proposed method is intuitively satisfactory, whereas the NN-based method

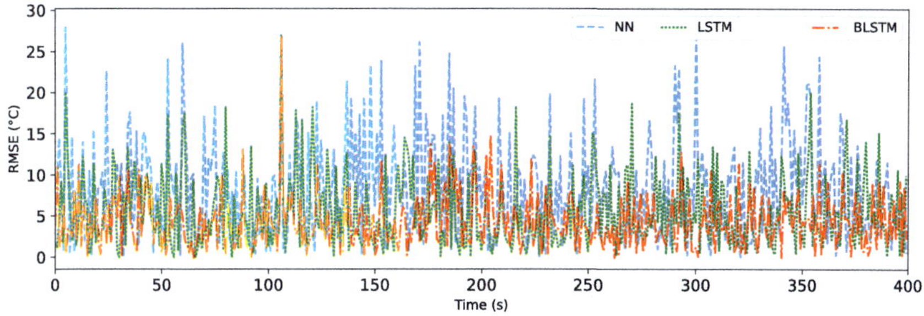

(a)

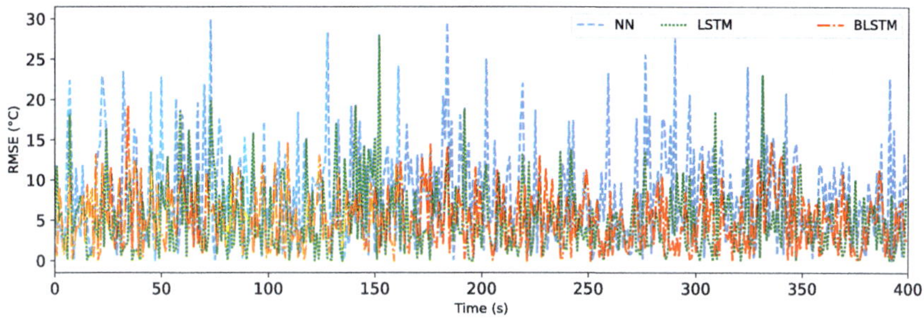

(b)

Figure 11.5 The error distributions over the whole validation prediction time domain, averaged over all the simulated cell temperature scenarios representing (a) Reactor #8889 and (b) Reactor #8888, respectively. The results by the NN and LSTM deterministic methods and Bayesian LSTM are also shown for comparison.

is less meaningful because of the many pulse errors found. In addition, evaluated from this subfigure, the averaged RMSE of the proposed method is also a boundary, which illustrates that the prediction performance is guaranteed under the credible prediction with uncertainty quantification.

As we can see, the proposed method outperforms other methods in terms of both forecasting accuracy and the dynamic error bound. The prediction MSE and RMSE by the proposed method are only 37.1418 and 4.8370 for reactor #8888, whereas they are 39.3373 and 5.0101 by the NN-based method and 38.4931 and 4.9414 by the LSTM. Compared to the NN-based method and LSTM, the proposed method achieves the lower RMSE and MSE, which means that the accuracy of the proposed Bayesian LSTM is better than its comparisons. More intuitively, in reactor #8888, the whole prediction error ratio (in terms of ε_1 and ε_2) by the proposed method is only 40.1‰ and 5.2‰, respectively, whereas for the competitors, by the NN-based method it is 42.4‰ and 5.4‰, which is

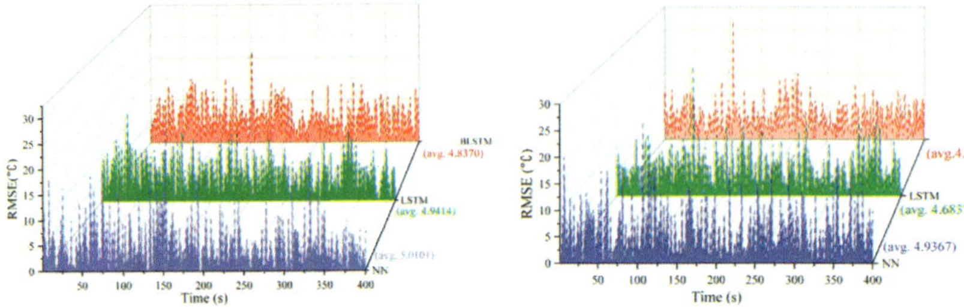

(a) (b)

Figure 11.6 The error distributions over the whole testing prediction time domain, averaged over all the simulated cell temperature scenarios representing (a) Reactor #8889 and (b) Reactor #8888, respectively. The results of the NN and LSTM deterministic methods and Bayesian LSTM are also shown for comparison.

higher than by the proposed approach. The LSTM performance in terms of ε_1 and ε_2 is 41.6‰, and 5.3‰, respectively, which is slightly higher than that of the Bayesian LSTM, which demonstrates that the time sequence network is very helpful for the dynamic cell temperature, and the historical observation is critical for the forecasting horizon information. As for reactor #8889, the RMSE and MSE indexes by the proposed method are 34.2009, and 4.6446 lower than by the NN-based method with 38.3346 and 4.9367 and by the deterministic LSTM, 34.7486 and 4.6837, respectively. On the other hand, the proposed method obtained the state-of-the-art ε_2 performance as 5.0‰, whereas the other two deterministic comparisons are 5.3‰ for NN and 5.1‰ for LSTM. Thus the results illustrate that, similar to the deterministic temperature prediction for the whole time period, the performance improvement by the proposed Bayesian LSTM is greater at reactor #8889 than at reactor #8888. In addition, attentively, the prediction error ratio in the testing dataset is up to 5.0‰ in terms of ε_2, which means that if the averaged temperature during aluminum electrolysis reduction cell in the forecasting horizon is 1000°C, then the prediction error is bounded within 5°C, and so this performance substantially outperforms previous works.

11.5 Conclusions

This framework proposes a novel aluminum temperature prediction algorithm based on probabilistic long short-term memory. The proposed framework uses the long short-term memory as the basic unit, combined with a probabilistic Bayesian network, so that the hidden features have a temporal instant scale. Furthermore, the prediction effectiveness and efficacy are also tested by the industrial aluminum reduction cell case. Compared to the NN-based and deterministic competitive prediction methods, the proposed probabilistic in two typical cases have superior performance and lower MSE. The evaluation indexes represent the final prediction error ratio bounded within 5‰ for reactor #8888 and 5.2‰ for the other reactor #8889, which is a great performance enhancement that is worth mentioning.

The proposed probabilistic LSTM-based prediction framework can be extended to other domains and improve the generalization of the prediction, especially some temporal time-scale systems with advanced control performance. The proposed method also achieves some similar prediction results to the generative adversarial network and Bayesian neural network, yet yields a more complex hidden representation. A possible direction to further explore our method is to generalize the process with faster sampling, which can be implemented by a class of non-Markovian processes. On the other hand, two different evaluation indexes are proposed to evaluate the temperature prediction method performance in a full dynamic time-scaled domain.

For future consideration, we will focus on the study of robust control with the model uncertainty, specifically, the local exploration and exploitation for the real-time switching of the different control methods, which could further improve the efficiency and performance of the prediction and control.

References

[1] Jie Wang, Yongfang Xie, Shiwen Xie, Xiaofang Chen, Optimization of aluminum fluoride addition in aluminum electrolysis process based on pruned sparse fuzzy neural network, ISA Transactions 133 (2023) 285–301.

[2] Zhaoke Huang, Chunhua Yang, Xiaojun Zhou, Weihua Gui, An improved TOPSIS-based multi-criteria decision-making approach for evaluating the working condition of the aluminum reduction cell, Engineering Applications of Artificial Intelligence 117 (2023) 105599.

[3] Yiyin Tang, Yalin Wang, Chenliang Liu, Xiaofeng Yuan, Kai Wang, Chunhua Yang, Semi-supervised LSTM with historical feature fusion attention for temporal sequence dynamic modeling in industrial processes, Engineering Applications of Artificial Intelligence 117 (2023) 105547.

[4] Yongfang Xie, Jinjing Yu, Shiwen Xie, Tingwen Huang, Weihua Gui, On-line prediction of ferrous ion concentration in goethite process based on self-adjusting structure RBF neural network, Neural Networks 116 (2019) 1–10.

[5] Keke Huang, Yiming Wu, Cheng Long, Hongquan Ji, Bei Sun, Xiaofang Chen, Chunhua Yang, Adaptive process monitoring via online dictionary learning and its industrial application, ISA Transactions 114 (2021) 399–412.

[6] Zhaohui Zeng, Weihua Gui, Xiaofang Chen, Yongfang Xie, Renchao Wu, A mechanism knowledge-driven method for identifying the pseudo dissolution hysteresis coefficient in the industrial aluminium electrolysis process, Control Engineering Practice 102 (2020) 104533.

[7] Yubo Sun, Xiaofang Chen, Weihua Gui, Lihui Cen, Yongfang Xie, Zhong Zou, Multi-scale 4D localized spatio-temporal graph convolutional networks for spatio-temporal sequences forecasting in aluminum electrolysis, Advanced Engineering Informatics 65 (2025) 103222.

[8] Keke Huang, Haofei Wen, Hongquan Ji, Lihui Cen, Xiaofang Chen, Chunhua Yang, Nonlinear process monitoring using kernel dictionary learning with application to aluminum electrolysis process, Control Engineering Practice 89 (2019) 94–102.

[9] Chunhua Yang, Longfei Zhou, Keke Huang, Hongquan Ji, Cheng Long, Xiaofang Chen, Yongfang Xie, Multimode process monitoring based on robust dictionary learning with application to aluminium electrolysis process, Neurocomputing 332 (2019) 305–319.

[10] Xiaoxue Wan, Lihui Cen, Xiaofang Chen, Yongfang Xie, Multi-generator adversarial dynamic spatial–temporal shapelet network for anode effect prediction in aluminum electrolysis process, Advanced Engineering Informatics 62 (2024) 102609.

[11] Xiaoxue Wan, Lihui Cen, Xiaofang Chen, Yongfang Xie, Weihua Gui, Prior knowledge-augmented unsupervised shapelet learning for unknown abnormal working condition discovery in industrial process, Advanced Engineering Informatics 60 (2024) 102429.

[12] Weichao Yue, Lingfeng Hou, Xiaoxue Wan, Yongfang Xie, Xiaofang Chen, Weihua Gui, Consensus-based probabilistic hesitant intuitionistic linguistic Petri nets for knowledge-intensive work of superheat degree identification, Advanced Engineering Informatics 59 (2024) 102261.

[13] Purushottam Gangsar, Rajiv Tiwari, Comparative investigation of vibration and current monitoring for prediction of mechanical and electrical faults in induction motor based on multiclass-support vector machine algorithms, Mechanical Systems and Signal Processing 94 (2017) 464–481.

[14] Yue Yu, Hamid Reza Karimi, Peiming Shi, Rongrong Peng, Shuai Zhao, A new multi-source information domain adaption network based on domain attributes and features transfer for cross-domain fault diagnosis, Mechanical Systems and Signal Processing 211 (2024) 111194.

[15] Zhaoke Huang, Chunhua Yang, Xiaofang Chen, Xiaojun Zhou, Weihua Gui, Time series clustering method with cluster validation to identify unknown local cell conditions in the aluminum reduction cell, Computers & Industrial Engineering 174 (2022) 108790.

[16] Zhaoke Huang, Chunhua Yang, Xiaofang Chen, Xiaojun Zhou, Guo Chen, Tingwen Huang, Weihua Gui, Functional deep echo state network improved by a bi-level optimization approach for multivariate time series classification, Applied Soft Computing 106 (2021) 107314.

[17] Yongxiang Lei, Xiaofang Chen, Ming Min, et al., A semi-supervised Laplacian extreme learning machine and feature fusion with CNN for industrial superheat identification, Neurocomputing 381 (2020) 186–195.

[18] Yongxiang Lei, Xiaofang Chen, Yongfang Xie, An improved cell situation identification approach with convolutional neural network and wavelet extreme learning machine, Proceedings of the Institution of Mechanical Engineers. Part I, Journal of Systems and Control Engineering 235 (10) (2021) 1898–1905.

[19] Yongxiang Lei, Fang Liu, Hamid Reza Karimi, Xiaofang Chen, Manifold semi-supervised learning for aluminum electrolysis temperature identification based on regularized hierarchical extreme learning machine, Proceedings of the Institution of Mechanical Engineers. Part I, Journal of Systems and Control Engineering 236 (6) (2022) 1109–1118.

[20] Yongxiang Lei, Hamid Reza Karimi, Xiaofang Chen, A novel self-supervised deep LSTM network for industrial temperature prediction in aluminum processes application, Neurocomputing 502 (2022) 177–185.

[21] Yongxiang Lei, Xiaofang Chen, Yongfang Xie, Lihui Cen, A self-supervised temporal temperature prediction method based on dilated contrastive learning, Journal of Process Control 120 (2022) 150–158.

[22] Bismark C. Torrico, Magno P. de Almeida Filho, Thiago A. Lima, Marcus D. do N. Forte, Rejane C. Sá, Fabrício G. Nogueira, Tuning of a dead-time compensator focusing on industrial processes, ISA Transactions 83 (2018) 189–198.

[23] Abdalrhman Abrahim Milad, Ibrahim Adwan, Sayf A. Majeed, Zubair Ahmed Memon, Munder Bilema, Hend Ali Omar, Maher G.M. Abdolrasol, Aliyu Usman, Nur Izzi Md Yusoff, Development of a hybrid machine learning model for asphalt pavement temperature prediction, IEEE Access 9 (2021) 158041–158056.

[24] Ning Wang, Jia-Yang Li, Efficient multi-channel thermal monitoring and temperature prediction based on improved linear regression, IEEE Transactions on Instrumentation and Measurement 71 (2021) 1–9.

[25] Zhiyu Sheng, Dong Wang, Jia Fu, Jinghua Hu, A computationally efficient spatial online temperature prediction method for PM machines, IEEE Transactions on Industrial Electronics 69 (11) (2021) 10904–10914.

[26] Yong Bao, Jiabin Jia, Online time-resolved reconstruction method for acoustic tomography system, IEEE Transactions on Instrumentation and Measurement 69 (7) (2019) 4033–4041.

[27] Dong Pan, Zhaohui Jiang, Weihua Gui, Ke Jiang, Xavier Maldague, Compensation method for the influence of dust in optical path on infrared temperature measurement, IEEE Transactions on Instrumentation and Measurement 70 (2020) 1–11.

[28] Dong Pan, Zhaohui Jiang, Weihua Gui, Xavier Maldague, Ke Jiang, Influence of dust on temperature measurement using infrared thermal imager, IEEE Sensors Journal 20 (6) (2019) 2911–2918.

[29] Karim Abdennadher, Pascal Venet, Gérard Rojat, Jean-Marie Rétif, Christophe Rosset, A real-time predictive-maintenance system of aluminum electrolytic capacitors used in uninterrupted power supplies, IEEE Transactions on Industry Applications 46 (4) (2010) 1644–1652.

[30] Zeng Shuiping, Wang Shasha, Qu Yaxing, Control of temperature and aluminum fluoride concentration based on model prediction in aluminum electrolysis, Advances in Materials Science and Engineering 2014 (2014).

[31] Tian Zhongda, Li Shujiang, Wang Yanhong, Wang Xiangdong, A multi-model fusion soft sensor modelling method and its application in rotary kiln calcination zone temperature prediction, Transactions of the Institute of Measurement and Control 38 (1) (2016) 110–124.

[32] Huixin Tian, Zhizhong Mao, Shu Wang, Kun Li, Application of genetic algorithm combined with BP neural network in soft sensor of Molten Steel temperature, in: 2006 6th World Congress on Intelligent Control and Automation, vol. 2, IEEE, 2006, pp. 7742–7745.

[33] Ling Shen, Zhipeng Chen, Zhaohui Jiang, Weihua Gui, Soft sensor modeling of blast furnace wall temperature based on temporal–spatial dimensional finite-element

extrapolation, IEEE Transactions on Instrumentation and Measurement 70 (2020) 1–14.

[34] Xiaofeng Yuan, Jiao Zhou, Biao Huang, Yalin Wang, Chunhua Yang, Weihua Gui, Hierarchical quality-relevant feature representation for soft sensor modeling: a novel deep learning strategy, IEEE Transactions on Industrial Informatics 16 (6) (2019) 3721–3730.

[35] Xiaofeng Yuan, Biao Huang, Yalin Wang, Chunhua Yang, Weihua Gui, Deep learning-based feature representation and its application for soft sensor modeling with variable-wise weighted SAE, IEEE Transactions on Industrial Informatics 14 (7) (2018) 3235–3243.

[36] Xiaofeng Yuan, Lin Li, Yuri A.W. Shardt, Yalin Wang, Chunhua Yang, Deep learning with spatiotemporal attention-based LSTM for industrial soft sensor model development, IEEE Transactions on Industrial Electronics 68 (5) (2020) 4404–4414.

[37] Chengzhu Wang, Yonggang Li, Keke Huang, Chunhua Yang, Weihua Gui, VAE4RSS: a VAE-based neural network approach for robust soft sensor with application to zinc roasting process, Engineering Applications of Artificial Intelligence 114 (2022) 105180.

[38] Michael Stalder, Firat Ozdemir, Artur Safin, Jonas Sukys, Damien Bouffard, Fernando Perez-Cruz, Probabilistic modeling of lake surface water temperature using a Bayesian spatio-temporal graph convolutional neural network, arXiv preprint, arXiv:2109. 13235, 2021.

[39] Xia Fang, Wei Zhang, Yuhao Guo, Jie Wang, Mei Wang, Shunlei Li, A novel reinforced deep RNN–LSTM algorithm: energy management forecasting case study, IEEE Transactions on Industrial Informatics 18 (8) (2021) 5698–5704.

[40] Juan Baz, Mikel Ferrero-Jaurrieta, Irene Díaz, Susana Montes, Gleb Beliakov, Humberto Bustince, Probabilistic study of induced ordered linear fusion operators for time series forecasting, Information Fusion 103 (2024) 102093.

[41] Jia Wang, Vincent W. Zheng, Zemin Liu, Kevin Chen-Chuan Chang, Topological recurrent neural network for diffusion prediction, in: 2017 IEEE International Conference on Data Mining (ICDM), IEEE, 2017, pp. 475–484.

[42] David Salinas, Valentin Flunkert, Jan Gasthaus, Tim Januschowski, DeepAR: probabilistic forecasting with autoregressive recurrent networks, International Journal of Forecasting 36 (3) (2020) 1181–1191.

[43] Xiaoning Zhang, Fang Fang, Jiaqi Wang, Probabilistic solar irradiation forecasting based on variational Bayesian inference with secure federated learning, IEEE Transactions on Industrial Informatics 17 (11) (2020) 7849–7859.

[44] Yandong Yang, Wei Li, T. Aaron Gulliver, Shufang Li, Bayesian deep learning-based probabilistic load forecasting in smart grids, IEEE Transactions on Industrial Informatics 16 (7) (2019) 4703–4713.

[45] Sarah Almaghrabi, Mashud Rana, Margaret Hamilton, Mohammad Saiedur Rahaman, Multivariate solar power time series forecasting using multilevel data fusion and deep neural networks, Information Fusion 104 (2024) 102180.

[46] Charles Blundell, Julien Cornebise, Koray Kavukcuoglu, Daan Wierstra, Weight uncertainty in neural network, in: International Conference on Machine Learning, PMLR, 2015, pp. 1613–1622.

[47] Yongxiang Lei, Hamid Reza Karimi, Lihui Cen, Xiaofang Chen, Yongfang Xie, Processes soft modeling based on stacked autoencoders and wavelet extreme learning machine for aluminum plant-wide application, Control Engineering Practice 108 (2021) 104706.

[48] Xiaogang Zhang, Yanying Lei, Hua Chen, Lei Zhang, Yicong Zhou, Multivariate time-series modeling for forecasting sintering temperature in rotary kilns using DCGNet, IEEE Transactions on Industrial Informatics 17 (7) (2020) 4635–4645.

CHAPTER 12

Direct data-driven quantile regressor forecastor for underflow concentration soft sensing: DDQRF

12.1 Introduction

The extraction of resources is a cornerstone of modern civilization, serving as a catalyst for societal progress, with the mining industry playing a central and indispensable role in this pursuit. Raw mining materials are essential to a wide range of applications, cutting across numerous sectors [1], including critical industries such as aerospace and medicine, further emphasizing the multifaceted importance of mining [2,3]. However, despite its pivotal contribution, the mining sector faces significant challenges. Chief among these is the precise regulation of underflow concentration, a critical quality parameter that directly influences the paste-filling process [4]. Accurate measurement of this concentration is hindered by several factors, including prohibitive device costs, complex coupling effects, and substantial time delays [5]. Maintaining a consistent and accurate underflow concentration is crucial for the robustness of paste-filling operations, as deviations can lead to imbalances and mud oscillations within thickeners. As a result, effective control of the cone thickener system (CTS) is vital to monitor paste extraction concentration and mitigate potential safety risks [6]. Given the mechanical complexities of CTS, reliable prediction of underflow concentration serves as a prerequisite for implementing effective control measures. Importantly, keeping the underflow concentration within a narrowly defined dynamic range is essential to prevent blockages or crushing while meeting the stringent quality demands of the paste-filling process.

The accurate underflow concentration is of critical importance in CTS. However, traditional time series forecasting models primarily focus on y_{t+1} subject to recent history $y : t = (y_t, \ldots, y_0)$ [7]. Among these, the Box–Jenkins methodology, including ARIMA models [8], is a

Learning-Based Soft Sensing and Predictions for Process Industries
https://doi.org/10.1016/B978-0-44-336759-5.00021-4

widely used point-estimation approach that employs historical data points in a recursive autoregressive framework for multi-horizon forecasting. Nonetheless, real-world forecasting often entails higher complexity, with time series exhibiting long-term dependencies between inputs and outputs [9]. Additionally, accurate estimation of prediction intervals is crucial for quantifying forecast uncertainties and optimizing control performance in decision-making, particularly for multistep, long-horizon predictions [10]. To tackle these challenges, a variety of modern techniques have been proposed, each addressing specific aspects of the forecasting problem [11].

Quantile uncertainty prediction serves as a pivotal approach in time series forecasting, particularly for applications requiring precise estimation of prediction intervals and robust decision-making. Unlike traditional point prediction methods that estimate the conditional mean of a target variable, quantile prediction focuses on estimating specific conditional quantiles of the target distribution. This approach provides a comprehensive representation of forecast uncertainties by generating prediction intervals at desired confidence levels. By learning to predict these quantiles directly, quantile regression eliminates the need for strong distributional assumptions, offering greater flexibility and applicability across diverse datasets [12]. Such predictions are particularly beneficial in scenarios with asymmetric costs associated with under- and over-prediction, as they enable more informed and risk-aware control strategies. Moreover, the ability to quantify uncertainty through prediction intervals enhances the reliability of forecasts in industrial systems, where maintaining operational consistency and safety is paramount. Consequently, quantile-based uncertainty prediction has emerged as a robust tool for advancing predictive analytics in complex, data-driven environments [13].

Most neural network-based approaches for time series forecasting adopt the recursive strategy [14], where the model predicts y_{t+1} based on history $y_{:t}$ and iteratively updates this estimate to forecast longer horizons. However, this recursive approach can suffer from error accumulation, particularly when employing recurrent neural networks (RNNs), as the model relies on its own predictions rather than on actual data during inference [15]. By contrast the direct strategy predicts y_{t+k} directly from $y_{:t}$ for each k, reducing bias, improving stability, and enhancing robustness against model misspecification [16]. Recent studies comparing multi-step forecasting strategies with neural networks highlight the efficacy of the direct multi-horizon approach, where the model is trained to predict a multivariate target $(y_{t+1}; \cdot; y_{t+k})$, thereby mitigating error accumulation while ensuring parameter efficiency. In decision-making contexts with asymmetric costs for over- and under-prediction, probabilistic forecast models that estimate the full conditional distribution $p(y_{t+k}|y_{:t})$ are preferable to point forecast models that predict only the conditional mean $E(y_{t+k}|y_{:t})$. Traditionally, for real-valued time series, probabilistic models assume Gaussian error distributions or stochastic processes for residual series $\varepsilon_t = y_t - \hat{y}_t$. However, quantile regression has gained prominence in scenarios where the prediction of specific quantiles is necessary for minimizing losses and quantifying uncertainties [12,13]. This method learns to predict conditional quantiles $y_{t+k}^{(q)}|y_{:t}$ of the target distribution without assuming specific distributional forms, enabling accurate probabilistic forecasts with precise prediction intervals. As a result, quantile regression is increasingly recognized as a robust approach for prediction calibration.

Direct data-driven control algorithms use input–output data to design control strategies, integrating parameter identification with model-based control. In industrial cone thickener systems (CTSs), several challenges underscore the need for a multi-horizon quantile strategy driven by direct data [6]. These challenges include:

i) The difficulty of measuring historical underflow concentration distributions, which limits the applicability of traditional Bayesian probabilistic predictors.

ii) The necessity for consistent operation and production in the paste-filling process, which demands accurate underflow concentration forecasts. Reliable forecasts enable precise control, enhance efficiency, and mitigate costs and safety hazards.

iii) The challenge of obtaining a comprehensive mathematical model of CTSs for implementing model-based methods, making data-driven approaches more viable. Recent advancements in direct data-driven prediction techniques offer promising solutions to address these challenges [17–19].

Based on the previous analysis, the time-series prediction of the CTS underflow concentration is critical yet remains insufficiently explored. In this work, we propose a consistent underflow concentration prediction method based on deep LSTM and quantile regression. The motivation is that the underflow concentration prediction is generated using a data-driven multi-horizon forecasting mechanism with uncertainty quantification. The key contributions of this chapter are:

- Novel architecture: This study proposes a new prediction model architecture (DDQRF) that implements direct data-driven multi-horizon prediction.
- Quantified uncertainty: The presented quantile regression model incorporates and quantifies prediction uncertainty.
- Learning capacity and robustness: The presented framework fully leverages the learning capacity of temporal machine learning models and the robustness of probabilistic models to uncertainty. Unlike traditional Bayesian probabilistic models, this study introduces a novel gating recurrent network (GRN) for encoder representation, characterized by its ease of use and implementation
- Industrial application: This study represents the inaugural effort to tackle the challenge of predicting underflow concentration within an industrial CTS system. The proposed model provides substantial insights, serving as a critical reference for control strategies.

12.2 Problem statement

In this section, we present a new sequence-to-sequence direct data-driven framework that generates multiple horizon quantile forecasts with uncertainty consideration for the industrial CTS. It is represented as

$$p(y_{t+k,i}, \cdots, y_{t+1,i} \mid y_{:t,i}, x^{(h)}_{:t,i}, x^{(f)}_{t:,i}, x^{(s)}_{i}), \tag{12.1}$$

where $y_{\cdot,i}$ represents the ith time series to be predicted, $x^{(h)}_{:t,i}$ denotes the historical temporal covariates, $x^{(f)}_{t:,i}$ contains the future information, and $x^{(s)}_{i}$ includes static, time-invariant features. Each time series serves as an individual input for a sequential neural network, such as an RNN or CNN.

In CTS, concentration is intertwined with numerous other variables [20]. In practice, the myriad of interdependent variables and the often opaque nature of their relationships with the target variable complicate the identification of relevant predictors. Furthermore, establishing the appropriate prediction horizon and the extent of historical observations required poses additional challenges. To address these complexities, a coefficient correlation analysis was conducted to identify key variables in the cone thickener production process. The selection of vital variables is a pivotal step in constructing a data-driven prediction model. Through mechanical analysis, the mud height and pressure were identified as significant factors due to their nonlinear correlations with the target variable and were thus incorporated into the model. Based on the foundation laid by [20], the input variables for the model include the mud pressure, mud height, most recent feed flow rate, flow rate of the highest volume, underflow density, and historical underflow concentration. Consequently, this sets the stage for multi-horizon prediction of underflow concentration in CTS systems:

$$\begin{aligned} & p(y_{t+k,i}, \cdots, y_{t+1,i} \mid y_{:t,i}, x_{:t,i}^{(h)}, x_{t:,i}^{(f)}, x_i^{(s)}) \\ & = p(y_{t+k,i}, \cdots, y_{t+1,i} \mid \langle \mathbf{C}_{i(t)}, \mathbf{Q}_{i(t)}, \mathbf{F}_{(t)}, \mathbf{C}_{o(t)}, \mathbf{Q}_{o(t)} \rangle) \\ & = p(y_{t+k,i}, \cdots, y_{t+1,i} \mid \mathbf{W} \cdot NN \sum_{i=1}^{L} \langle y_i, y_i' \rangle, \end{aligned} \tag{12.2}$$

where NN represents the neural network-based predictor, the corresponding weights are represented by $\mathbf{W}$, and the input variables are represented by the matrix $[C_i, Q_i, F, C_o, Q_o]$. However, prediction intervals are very insightful for the control and operation. So, Eq. (12.2) with quantile addition can be rewritten as

$$\begin{aligned} & p(y_{t+k,i}, \cdots, y_{t+1,i} \mid y_{:t,i}, x_{:t,i}^{(h)}, x_{t:,i}^{(f)}, x_i^{(s)}, q) \\ & = p(y_{t+k,i}, \cdots, y_{t+1,i} \mid \langle \mathbf{C}_{i(t)}, \mathbf{Q}_{i(t)}, \mathbf{F}_{(t)}, \mathbf{C}_{o(t)}, \mathbf{Q}_{o(t)} \rangle, q) \\ & = p(y_{t+k,i}, \cdots, y_{t+1,i} \mid \mathbf{W} \cdot NN \sum_{i=1}^{L} \langle y_i, y_i' \rangle, \end{aligned} \tag{12.3}$$

which implements the predicted qth quantile of the k-step-ahead forecast at instant t.

12.3 Main framework

DDQRF considers the uncertainty. As shown in Fig. 13.2, the original data is pre-processed into three subsequent data, the static metadata, past inputs, and the known future inputs. The static metadata is calculated by

$$\mathbf{c} = Cov(X_{L+k}, \ldots, X_{2L+k}) = E[(X_i - \mu_{X_i})(X_j - \mu_{X_j})], \; i \neq j, \tag{12.4}$$

where μ_{X_i} and μ_{X_j} are the expectations for the static metadata.

The past inputs and future inputs are both transformed to the LSTM, inherited by the gate mechanism and normalization term, and then the output of the first LSTM encoder would be

$$\phi(X, c) = LSTM(W, X, c). \tag{12.5}$$

Then the final output after the gate mechanism and layer normalization is described as

$$\mathbf{a} = LayerNorm(GLU(\phi(X, c))). \tag{12.6}$$

Gated recurrent network: To give the model the flexibility to apply non-linear processing only when needed and also to avoid the gradient varnishing issues, we propose a new architecture as the building block of the framework. The GRN takes the secondary inputs extracted from the LSTM encoders **a** and static covariate **c** and then yields

$$\begin{aligned} GRN(\mathbf{a}, \mathbf{c}) &= Norm(\mathbf{a} + GLU(\eta_1)), \\ \eta_1 &= \mathbf{W}_1\eta_2 + \mathbf{b}_1, \\ \eta_2 &= ELU(\mathbf{W}_2\mathbf{a} + \mathbf{W}_3\mathbf{c} + \mathbf{b}_2), \\ GLU(\gamma) &= \sigma(\mathbf{W}_4\gamma + \mathbf{b}_4) \odot (\mathbf{W}_5\gamma + \mathbf{b}_5), \end{aligned} \tag{12.7}$$

where ELU is the exponential linear unit activation kernel, and **b** is the bias of the ELU.

As presented in the previous works, the underflow concentration is continuous and linked with the previous feature representation. In the designed framework, the decoder representation is given as

$$\varphi(X, a, c) = LayerNorm(GLU(\xi)). \tag{12.8}$$

The decoder representation gives the high-level features of the underflow concentration prediction. In the top layer, the model generates the prediction intervals by various simultaneous quantiles. Probabilistic quantile forecasters are activated by the linear transformation of the output from the temporal decoder representation:

$$\hat{y}(q, t, k) = W_q\hat{\varphi}(X, a, c) + b_q, \tag{12.9}$$

where k is the horizon in the future, $W_q \in \mathbb{R}^{1\times L}$, and $b_q \in \mathbb{R}$ is the bias for quantile q. In quantile regression, models are trained to minimize the total quantile loss $\mathcal{L}_q$:

$$\mathcal{L}_q(y, \hat{y}, q) = q(y - \hat{y})_+ + (1 - q)(\hat{y} - y)_+, \tag{12.10}$$

where $(\cdot)_+ = \max(0, \cdot)$. When $q = 0.5$, $\mathcal{L}_q$ is simply the mean absolute error (MAE), and its minimizer is the median of the predictive distribution. Let τ be the maximum number of forecast horizons, and let $\mathcal{M}$ be the matrix of quantiles of interest. Then the $\tau \times \mathcal{M}$ matrix $\hat{\mathbf{Y}} = \left[\hat{y}_{t+k}^{(q)}\right]_{k,q}$ is the output of a parametric model $g(y_{:t}, x, \theta)$, for example, an RNN. The model parameters are trained to minimize the total loss $\sum_t \sum_q \sum_k \mathcal{L}_q(y_{t+k}, \hat{y}_{t+k}^{(q)})$, where t iterates through all forecast creation times (FCTs). Depending on the problem, the components of the sum can be assigned different weights, to highlight or discount different quantiles and horizons. In the underflow concentration prediction, the loss function would be the joint quantile loss, which is summed over all the quantile outputs:

$$\mathcal{L}(\mathcal{N}, \mathbf{W}) = \sum_{y_t \in \mathcal{N}} \sum_{q \in \mathcal{M}} \sum_{k=1}^{k=\tau} \frac{\mathcal{L}_q(y, \hat{y}, q)}{\mathcal{N}k}, \tag{12.11}$$

where $\mathcal{N}$ is the domain field of all the data-driven training data, $\mathbf{W}$ is the weight of the proposed model, $\mathcal{M}$ is the quantile hyper-parameter list that is denoted manually; in our experiment, $\mathcal{M} = [0.1, 0.5, 0.9]$. Finally, the predicted framework presented is formulated, and an intuitive demonstration is given in Fig. 12.1.

In this framework, the inputs are static metadata, time-varying past inputs, and time-varying a priori known future. Variable selection is used for judicious selection of the most salient features based on the process key quality variables. Gated residual network blocks enable efficient information flow with skip connections and gating layers. LSTM mechanism is integrated as the encoder representation for the robust feature aggregation, and the output for the prediction is obtained from any time-dependent quantile regression multi-horizon forecaster.

12.4 Performance evaluation

To fulfill the validation and verification process, the LSTM and GRN are selected as the comparison benchmarks. The ablation study results are summarized in Fig. 12.2. Fig. 12.2a compares the different regressors with different quantiles. The observation underflow concentration is fully bounded by the upper and lower boundaries, Regressor 1 (green dash line) and Regressor 3 (the dotted line). It demonstrates that the proposed framework correctly predicts the underflow concentration in

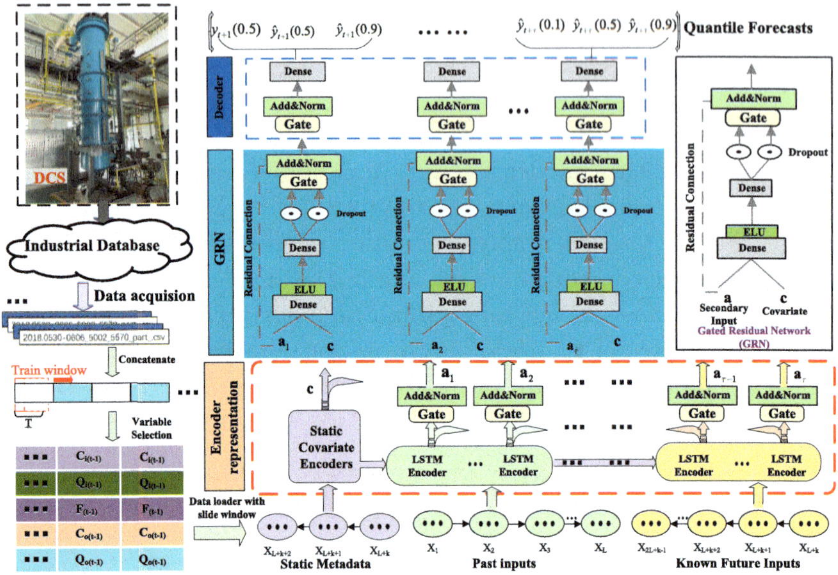

Figure 12.1 Flowchart of the proposed direct data-driven underflow concentration prediction method.

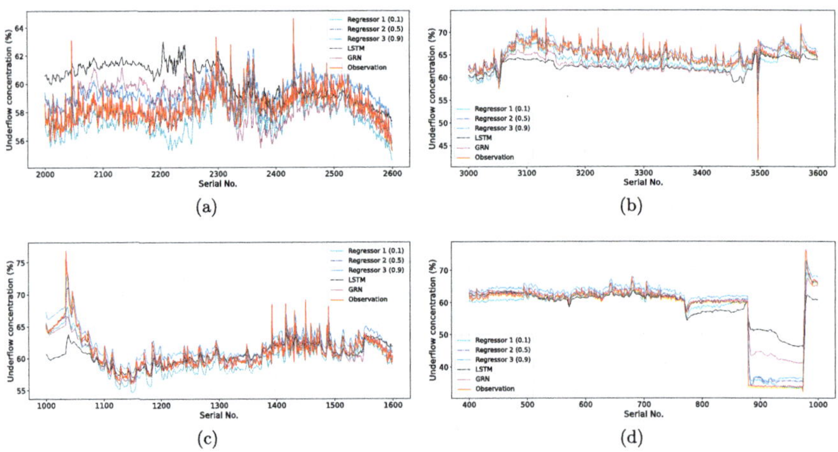

Figure 12.2 The proposed model for industrial underflow prediction results in training, validation, and testing data.

multiple horizons. On the other hand, for Regressor 2 (the blue dot line), also the mean probabilistic prediction follows the actual underflow concentration value timidly, which means that the proposed method

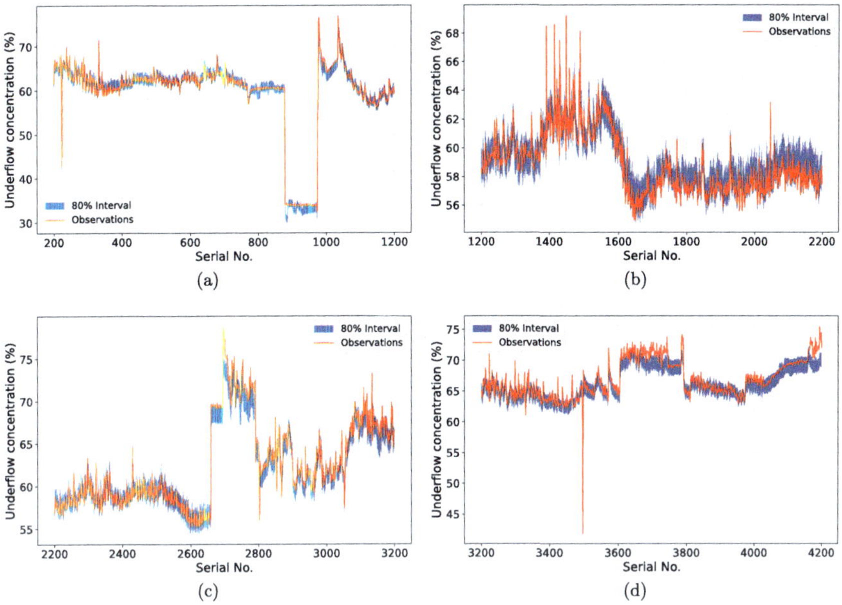

Figure 12.3 The proposed GRU model for industrial underflow prediction results. The subfigure is a different time series, and all the results are conducted on the testing metadata.

achieves a great performance. Similar characteristics are also discovered in the subsequent series; the presented method implements a satisfactory prediction performance. Compared to the presented approach, the LSTM, the black line in Fig. 12.2a, shows an apparent residual error between the prediction and observation. Fig. 12.2b shows that the GRN and LSTM usually underestimate the underflow concentration, whereas the proposed Regressor 2 showed a superior performance. Fig. 12.2c also shows that in time instant $t = 1000$, the actual underflow concentration is around 65%, whereas the LSTM predicted value is 60%, such a prediction residual degrades the LSTM model performance. Fig. 12.2b verified the initial performance of those different methods; as shown in the interval [880, 980], the proposed three quantile regressors achieve accurate prediction, whereas the LSTM and GRN caused inaccurate underflow concentration prediction, with almost 15% and 10% residual errors, respectively, see Fig. 12.4.

As for the verification, the presented method is deployed to the CTS. Fig. 12.3 presents the application performance of the proposed quantile

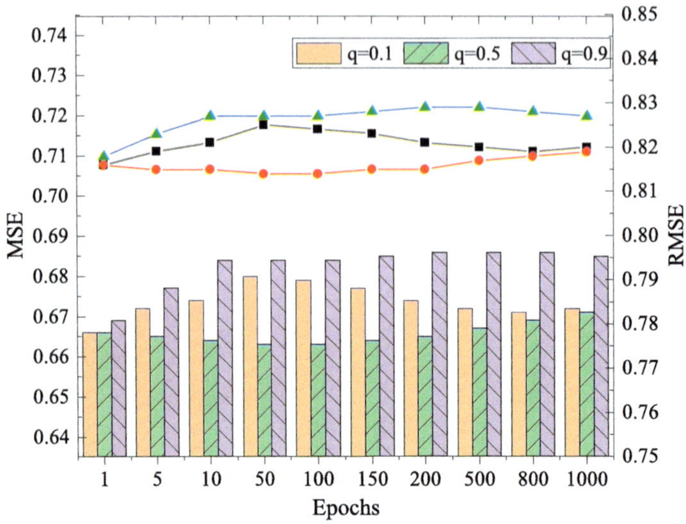

Figure 12.4 The error performance comparison (RMSE and MSE) in an increasing epoch.

forecaster. Figs. 12.3a–12.3d describe four different industrial applicative series. The main feature is that all the observations are bounded by an 80% prediction interval, which is computed by the prediction upper boundary ($q = 0.9$) and lower boundary ($q = 0.1$). The proposed regressor covers the observation very well, even in some abrupt oscillation intervals, such as [900, 1000]. However, the proposed method also shows a weakness, which is the ability to cucumber the situation of a single outlier; the intuitive point is shown in Fig. 12.3a around instant $t = 220$ and in Fig. 12.3d around instant $t = 3500$. This result is reasonable because the prediction accuracy of the proposed method is higher than other state-of-the-art counterparts, whereas other methods have a worse ability to deal with the outlier. Further, in the practical deployment, the single outlier is not the main factor that should be prioritized.

The more detailed ablation study results are given in Fig. 12.4. An increasing epoch with different quantiles is conducted while keeping other hyper-parameter fixed. Fig. 12.4 gives the intuitive result presentation of the MSE, associated with the bar, and RMSE with lines. Compared to $q = 0.1$ and $q = 0.9$, performance with $q = 0.5$ outperforms others, with the lowest MSE and RMSE, see Table 12.1. It also shows that when the epoch is set to 100, associated with quantile coefficient $q = 0.5$, we obtain the lowest MSE 0.662. Therefore, as fully explored, for the deployment

Table 12.1 Evaluation indexes for the different competitive prediction methods. Compared to the existing state-of-the-art deterministic approaches, our proposed benchmark implements probabilistic data-driven forecasting, and also multiple horizon prediction.

Categories	Methods	RMSE	MAE
Deterministic	RNN	0.6712 ± 0.0124	0.3488 ± 0.0082
	GRN	0.4634 ± 0.0078	0.2499 ± 0.0345
	LSTM	0.7533 ± 0.0030	0.5119 ± 0.0079
	SVR	0.8133 ± 0.0002	0.6003 ± 0.0013
	BiLSTM	0.8091 ± 0.0092	0.6083 ± 0.0118
	GSTAE	0.7735 ± 0.0068	0.5296 ± 0.0097
	SS-PdeepFM	0.7650 ± 0.0036	0.5206 ± 0.0073
Probabilistic	DualLSTM	0.7985 ± 0.0116	0.5847 ± 0.0164
	DeepAR	0.7571 ± 0.0033	0.5235 ± 0.0092
	Suggested	$\mathbf{0.6654 \pm 0.0021}$	$\mathbf{0.631 \pm 0.0013}$
Recursive point strategy	ARIMA	0.7515 ± 0.0034	0.2074 ± 0.0072
	LSTM-DeepFM	0.7497 ± 0.0026	0.5091 ± 0.0071
	MPA-RNN	0.7515 ± 0.0034	0.2074 ± 0.0072
	DA-RNN	0.851 ± 0.0029	0.2322 ± 0.0052
	DA-LSTM	0.851 ± 0.0029	0.2322 ± 0.0052
	Attention	0.9086 ± 0.0021	0.2445 ± 0.0031
Multiple-horizon strategy	DeepAR	0.9086 ± 0.0021	0.2445 ± 0.0031
	DSTP-RNN	0.8496 ± 0.0051	0.2254 ± 0.0042
	Suggested	$\mathbf{0.6654 \pm 0.0021}$	$\mathbf{0.631 \pm 0.0013}$

verification, the optimal epoch is set to 100. When the epoch is set to 50, $q = 0.1$ and $q = 0.9$ achieve the highest RMSE, 0.824 and 0.827, respectively, whereas for $q = 0.5$, it is 0.814. The result explores the optimal hyper-parameters for the proposed framework.

Table 12.1 gives the comparison of different competitors, the deterministic methods SVR [21], MPA-RNN [22], SS-PdeepFM [23], and DSTP-RNN [24], the probabilistic approaches DualLSTM [25] and DeepAR [26], and the recursive point strategies DA-RNN [27] and DA-LSTM [28]. The proposed method is a scope of multi-horizon and probabilistic. Compared to the basic LSTM, the GRN improves the prediction ability, whereas the GRN feature learning ability is also limited. Therefore the proposed method has a lower MSE and RMSE, compared to its own basic benchmarks. Despite the single feature, the proposed method integrates these advantages.

12.5 Conclusions

This work provides a high-efficiency key-quality underflow prediction model for the deep cone thickener system to address the challenges of underflow prediction in industrial CTS plants. It presents a novel probability underflow concentration soft sensing method for industrial paste filling process. Compared to the existing deterministic methods, we first implement probabilistic underflow concentration forecasting. Besides, differing from the probabilistic Bayesian network, a new quantile regression with a multiple-horizon forecasting model is proposed. This framework gives a benchmark to other similar process industries, which can easily be employed and used. The proposed method can take full advantage of the time sequence model's nonlinear mapping ability, and the probabilistic model's uncertainty quantification. The validation and verification experiments in the industrial CTS demonstrated that the performance is superior to other counterparts. For further research, we will mainly focus on the following aspect: As discussed in the previous figure, there are some outliers that made the regressor unstable. Some algorithms that smooth the outlier can be further explored and employed in the proposed probabilistic regression forecaster.

References

[1] Omid Bahrami, Wentao Wang, Rui Hou, Jerome P. Lynch, A sequence-to-sequence model for joint bridge response forecasting, Mechanical Systems and Signal Processing 203 (2023) 110690.

[2] Aixiang Wu, Yong Wang, Zhu'en Ruan, Bolin Xiao, Jiandong Wang, Linqi Wang, Key theory and technology of cemented paste backfill for green mining of metal mines, Green and Smart Mining Engineering 1 (1) (2024) 27–39.

[3] Lei Bo, Shangqing Yang, Yang Liu, Zihang Zhang, Yiying Wang, Yanwen Wang, Coal mine solid waste backfill process in China: current status and challenges, Sustainability 15 (18) (2023) 13489.

[4] Yongxiang Lei, Hamid Reza Karimi, Underflow concentration prediction based on improved dual bidirectional LSTM for hierarchical cone thickener system, The International Journal of Advanced Manufacturing Technology (2023) 1–12.

[5] Dong Xiao, Hongfei Xie, Longqiang Jiang, Ba Tuan Le, Jichun Wang, Chongmin Liu, Hongzong Li, Research on a method for predicting the underflow concentration of a thickener based on the hybrid model, Engineering Applications of Computational Fluid Mechanics 14 (2020) 13–26, https://doi.org/10.1080/19942060.2019.1658228.

[6] Zhaolin Yuan, Jinlong Hu, Di Wu, Xiaojuan Ban, A dual-attention recurrent neural network method for deep cone thickener underflow concentration prediction, Sensors 20 (5) (2020) 1260.
[7] Dario Valcamonico, Piero Baraldi, Enrico Zio, Luca Decarli, Anna Crivellari, Laura La Rosa, Combining natural language processing and Bayesian networks for the probabilistic estimation of the severity of process safety events in hydrocarbon production assets, Reliability Engineering & System Safety 241 (2024) 109638.
[8] George E.P. Box, Gwilym M. Jenkins, Gregory C. Reinsel, Greta M. Ljung, Time Series Analysis: Forecasting and Control, John Wiley & Sons, 2015.
[9] Max Jaderberg, Karen Simonyan, Andrew Zisserman, et al., Spatial transformer networks, Advances in Neural Information Processing Systems 28 (2015).
[10] Léonard Torossian, Victor Picheny, Robert Faivre, Aurélien Garivier, A review on quantile regression for stochastic computer experiments, Reliability Engineering & System Safety 201 (2020) 106858.
[11] Lin-Feng Mei, Wang-Ji Yan, Ka-Veng Yuen, Wei-Xin Ren, Michael Beer, Transmissibility-based damage detection with hierarchical clustering enhanced by multivariate probabilistic distance accommodating uncertainty and correlation, Mechanical Systems and Signal Processing 203 (2023) 110702.
[12] James W. Taylor, A quantile regression neural network approach to estimating the conditional density of multiperiod returns, Journal of Forecasting 19 (4) (2000) 299–311.
[13] Sheng Xiang, Yi Qin, Jun Luo, Fei Wu, Konstantinos Gryllias, A concise self-adapting deep learning network for machine remaining useful life prediction, Mechanical Systems and Signal Processing 191 (2023) 110187.
[14] Daoguang Yang, Hamid Reza Karimi, Kangkang Sun, Residual wide-kernel deep convolutional auto-encoder for intelligent rotating machinery fault diagnosis with limited samples, Neural Networks 141 (2021) 133–144.
[15] Ying Zhang, Ming Zhang, Chao Liu, Zhipeng Feng, Yuchun Xu, Reliability enhancement of state of health assessment model of lithium-ion battery considering the uncertainty with quantile distribution of deep features, Reliability Engineering & System Safety (2024) 110002.
[16] Amandine Marrel, Bertrand Iooss, Probabilistic surrogate modeling by Gaussian process: a review on recent insights in estimation and validation, Reliability Engineering & System Safety (2024) 110094.
[17] Yongxiang Lei, Intelligent optimal framework for the industrial mining plant-wide prediction control, IEEE Transactions on Instrumentation and Measurement (2023).
[18] Ahmed Mosallam, Kamal Medjaher, Noureddine Zerhouni, Data-driven prognostic method based on Bayesian approaches for direct remaining useful life prediction, Journal of Intelligent Manufacturing 27 (2016) 1037–1048.
[19] Kailong Liu, Yunlong Shang, Quan Ouyang, Widanalage Dhammika Widanage, A data-driven approach with uncertainty quantification for predicting future capacities and remaining useful life of lithium-ion battery, IEEE Transactions on Industrial Electronics 68 (4) (2020) 3170–3180.
[20] Yongxiang Lei, Hamid Reza Karimi, Lihui Cen, Xiaofang Chen, Yongfang Xie, Processes soft modeling based on stacked autoencoders and wavelet extreme learning machine for aluminum plant-wide application, Control Engineering Practice 108 (2021) 104706.
[21] Xiaofeng Yuan, Shuaibin Qi, Yalin Wang, Haibing Xia, A dynamic CNN for nonlinear dynamic feature learning in soft sensor modeling of industrial process data, Control Engineering Practice 104 (2020) 104614.
[22] J. Geng, C. Yang, Y. Li, et al., MPA-RNN: a novel attention-based recurrent neural networks for total nitrogen prediction, IEEE Transactions on Industrial Informatics 18 (10) (2022) 6516–6525.

[23] Lei Ren, Tao Wang, Yuanjun Laili, Lin Zhang, A data-driven self-supervised LSTM-DeepFM model for industrial soft sensor, IEEE Transactions on Industrial Informatics 18 (9) (2021) 5859–5869.
[24] Yeqi Liu, Chuanyang Gong, Ling Yang, Yingyi Chen, DSTP-RNN: a dual-stage two-phase attention-based recurrent neural network for long-term and multivariate time series prediction, Expert Systems with Applications 143 (2020) 113082.
[25] Yongxiang Lei, Hamid Reza Karimi, DualLSTM: a novel key-quality prediction for a hierarchical cone thickener, Control Engineering Practice 137 (2023) 105566.
[26] David Salinas, Valentin Flunkert, Jan Gasthaus, Tim Januschowski, DeepAR: probabilistic forecasting with autoregressive recurrent networks, International Journal of Forecasting 36 (3) (2020) 1181–1191.
[27] Yao Qin, Dongjin Song, Haifeng Chen, Wei Cheng, Guofei Jiang, Garrison Cottrell, A dual-stage attention-based recurrent neural network for time series prediction, arXiv preprint, arXiv:1704.02971, 2017.
[28] Ye Zhu, Shiwen Xie, Yongfang Xie, Xiaofang Chen, Temperature prediction of aluminum reduction cell based on integration of dual attention LSTM for non-stationary sub-sequence and ARMA for stationary sub-sequences, Control Engineering Practice 138 (2023) 105567.

CHAPTER 13

A novel key-quality prediction framework for industrial deep cone thickener: DualLSTM

13.1 Introduction

Resource mining is a critical industry for societal development and progress. Raw mining products have been widely applied to many domains, such as aerospace and medicine. The hierarchical cone thickener is a critical device for paste filling in the mining industry [1,2]. The current mining and paste-filling industry still faces some thorny challenges. For example, the underflow concentration of the cone thickener system (CTS) directly determines the quality of the whole pasting process [3]. The underflow concentration of CTS is a critical key quality variable, which is hard to measure because of the multi-couples, high device cost, and large time delay [4]. An accurate stable underflow concentration ensures the stability of the paste-fill process. An oscillating underflow concentration would cause an imbalance and mud oscillation in the thickener. In this large-scale device, the prior task is to control the underflow concentration in the CTS. However, the CTS is a large time-delay process in that the underflow concentration needs to be increased by the paste gravity filling. The filling quality largely relies on the underflow concentration of the CTS. The core requirement of the pump filling is that it should keep the underflow concentration in a small dynamic range. If the underflow concentration is too high, then it may cause filling accidents such as blockage and crushing. If the underflow concentration is too low, then it cannot meet the requirement of paste filling and results in safety hazards. Therefore taking into account the mechanical process of CTS and then accurately predicting the underflow concentration serve as a basis for further control purposes, maintaining the quality of paste mining filling and avoiding safety hazards [5].

The traditional physical device can hardly achieve the online long underflow concentration prediction because the current CTS is a multi-coupled system. The different sampling frequency of pressure and flow

Learning-Based Soft Sensing and Predictions for Process Industries
https://doi.org/10.1016/B978-0-44-336759-5.00022-6

leverages the massive integrated sensors to detect and monitor the production process. Furthermore, much historical information and factors among the quality variables should be considered. Based on the industrial case analysis, some solutions to these problems have been proposed. To sum up, the existing methods can mainly be classified into two categories: 1) the mathematical model-centric mechanism (also called the "white box") and 2) the model-free data-centric methods ("black box"). In the former case, however, the performance of white box-based models is determined by the specific mathematical model of the device, which means that the accuracy will greatly decrease if the model is incorrect. Furthermore, there are a lot of devices that do not have a mechanical mathematical model, so these models lose their significance. As for the latter case, with the development of machine learning, many data-driven self-supervised and semi-supervised methods have been proposed in the literature to improve the modeling of industrial processes [6,7]. The main advantage of these methods is that they can learn a nonlinear mapping projection between the input variables and target outputs. Compared to traditional physical sensor devices [8–17], it is the most economical way to measure key quality, which can save costs and increase profits.

Over the last decades, deep learning technology has become a hot topic in various research fields. For some time-series tasks, many temporal models have emerged, for example, recurrent neural network (RNN) [18], long short-term memory (LSTM) [19], gated recurrent unit (GRU) [20], Transformer [21], feature representation [22], etc. In [23] the time series factors have been considered for many plant-wide applications [24]. For instance, some improved attention mechanism has been applied to the models [25]. However, in the specific CTS system, some research has been put into the deep cone thickener systems, and the time instant key factor has been ignored. For example, they proposed an underflow concentration prediction method based on the XGBoost model, which achieves satisfactory performance in the underflow concentration of CTS. Nevertheless, the results in [5] are based on a static prediction method, and the historical and temporal information has been lost during the network training stages. The information contained in time-instant factors has been underestimated. Recently, a dual-attention RNN method is used for the deep cone thickener application; their prediction approach considers the temporal pertain to the same quality factors and achieves comparative accuracy and efficiency, whereas its performance relies on fine-tuning the hyper-parameters, and the generation ability has been greatly decreased. In [5] a spatiotemporal attention-based LSTM for industrial soft sensor model is introduced, which achieves great development in the processes industry. In [26] an automatic identification system (AIS) data-driven LSTM method is investigated based on the fusion of the forward sub-network and the reverse sub-network (termed as FRA-LSTM) to predict the vessel trajectory; however, their methods did not consider the noise processing and time-spatial factors. In [27] a novel attention-based recurrent neural network for nitrogen prediction is proposed, where three kinds of attention mechanisms are used in the MPA-RNN model. In [28] the specific investigation of the forward and backward gradient of the LSTM unit is reported, which covers the implementation of the training step with backpropagation through time. Some other variants of the recent LSTM-based model can be found in [29,8,16,30]. Compared to the traditional industry [31,6], the data collected directly by the exogenous infrequent sensor in CTS have large disturbances, errors, and occasional sudden outliers. In the proposed framework, this problem can be alleviated through appropriate filtering.

Based on the aforementioned analysis, the time series prediction of the underflow concentration of CTS has not been yet completely studied. Developed from [5], a novel underflow concentration method based on deep LSTM and average smooth filtering is proposed. The motivation is that the average smooth filtering is used to prepossess the industrial CTS data, and then a dual bidirectional LSTM (BiLSTM) with that integrated attention mechanism is generated for the underflow concentration prediction. The main solid bullet benefits of this chapter are as follows.

- Unlike the traditional RNN, LSTM, and bidirectional LSTM, a new prediction model architecture (DualLSTM) is proposed. The proposed architecture uses the feedforward and reverse BiLSTM as the basic units, which implement the new information learning. Average smoothing filtering is considered in the plant-wide process and thus Gaussian white noise cannot affect the performance of the proposed model. The robustness and generalization ability have been verified. The prediction accuracy is improved by a large margin compared to the existing state-of-the-art methods.
- Distinct from the previous prediction method in [32,29,5,33], a novel back-propagation through time (BPTT) algorithm is designed for the DualLSTM. The proposed solution for the new method is another key contribution of this brief.

- To the best of our knowledge, we first employ the DualLSTM to tackle the underflow concentration prediction in industrial CTS. The proposed model can be a core part of further intelligent control, which is the basis for further optimization control in an integrated platform.

To verify the feasibility and effectiveness of the proposed method, a realistic industrial application in Beijing, China, is covered by the core DualLSTM verification.

13.2 Problem statement

For subsurface paste filling, a deep cone thickener is essential. The diagram of CTS is shown in Chapter 4. CTS is a key method for achieving a consistent concentration for underground mining fills. Pipe-blocking mishaps may occur during the thickening process if the subterranean concentration is too high. On the other hand, a low subterranean concentration reduces the quality of the entire backfilled paste and, as a result, the overall safety of the mining operation. Therefore developing a model to forecast subterranean concentrations in the CTS system is critical. The whole mining paste filling operates on a continuous hierarchical concept. The CTS was fed with the crude unstable low-concentration slurry flow (almost 20–30%), which was combined with a flocculant to speed up the sinking rate. A mud bed can collect dissolved particles. The suitable concentration and volume feed flow are created at the bottom of the CTS. The top clean water from the overflow pipe has also been recycled for future use. The CTS's primary control needs to provide a consistent and precise underflow concentration. The concentration of underflow is a key metric for assessing the effectiveness and efficiency of the industrial underground pasting process. It is necessary to identify changes in the different variables due to the inner link for the production quality variable in the deep cone thickening process. This time-series architecture may be used to acquire some prior knowledge and historical information, which can subsequently be used to anticipate the underflow concentration. The specific statement of the problem is

Given: $(\mathbf{x}_1, \mathbf{x}_2, \ldots, \mathbf{x}_n, \mathbf{Y})$

Predict: $\widehat{\mathbf{Y}} = \left(y_{T+1}, y_{T+2}, \ldots, y_{T+t}\right)^{\mathrm{T}} \in \mathbb{R}^t$,

where $(\mathbf{x}_1, \mathbf{x}_2, \ldots, \mathbf{x}_n, \mathbf{Y})$ is the sequential and historical quality variables series, and T denotes the instant of the underflow concentration.

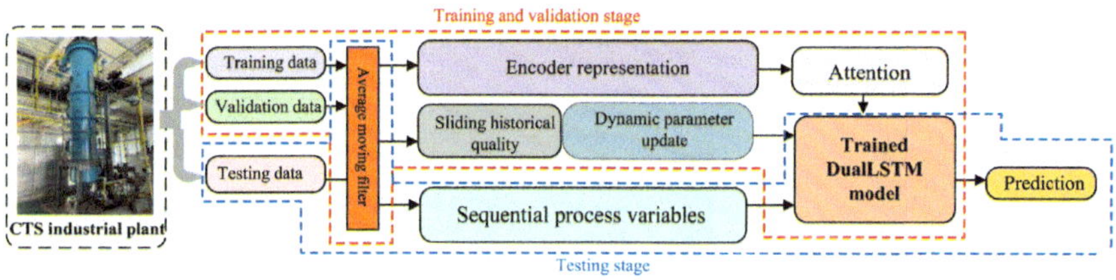

Figure 13.1 The flowchart illustration of DualLSTM.

Specifically, we need to learn a nonlinear mapping representation with $(\hat{y}_{T+1}, \hat{y}_{T+2}, \ldots, \hat{y}_{T+\tau}) = F(y_1, \ldots, y_T, x_1, \ldots, x_n)$.

13.3 Main framework

The framework and flowchart of the whole proposed algorithms are given in Fig. 13.1. Two stages of mechanisms are used for training. The first stage is coordinate training, which means that all the historical variable data are used to train the DualLSTM model for acquiring the optimal parameters. The solution process is fine-tuned with a background gradient algorithm. The backpropagation through time (BPTT) [29] is used for training. With BPTT, the maximum number of instant steps is set along with errors that can be propagated, see Fig. 13.2.

In the proposed framework, $h_{(t-1)}$ is the hidden state, $x_{(t)}$ is the input variable vector, and $T_{(t)}$ is the prediction time. Let the key variables for training and underflow concentration sequences be $\{x_{(1)}, x_{(2)}, \ldots, x_{(N)}\}$ and $\{T_{(1)}, T_{(2)}, \ldots, T_{(N)}\}$, where N is the total number of training data in the time sequence. Suppose that the intended behavior of the learning process is to observe the memory gate state at the end of the epoch $[t, t+s]$. The desired underflow concentration or the output loss is

$$L^{(t)}(y, \hat{y}) = \frac{1}{N}\sum_{i=1}^{N}\left(y_{(t)} - \hat{y}_{(t)}\right)^2. \tag{13.1}$$

The next target is to train the whole network parameters. Consider the dual Bi-LSTM having two embeddings with feedforward LSTM and reverse LSTM. Thus, developed from [28], the DualLSTM fine-tuning policy for the hyper-parameters is designed in the sequel.

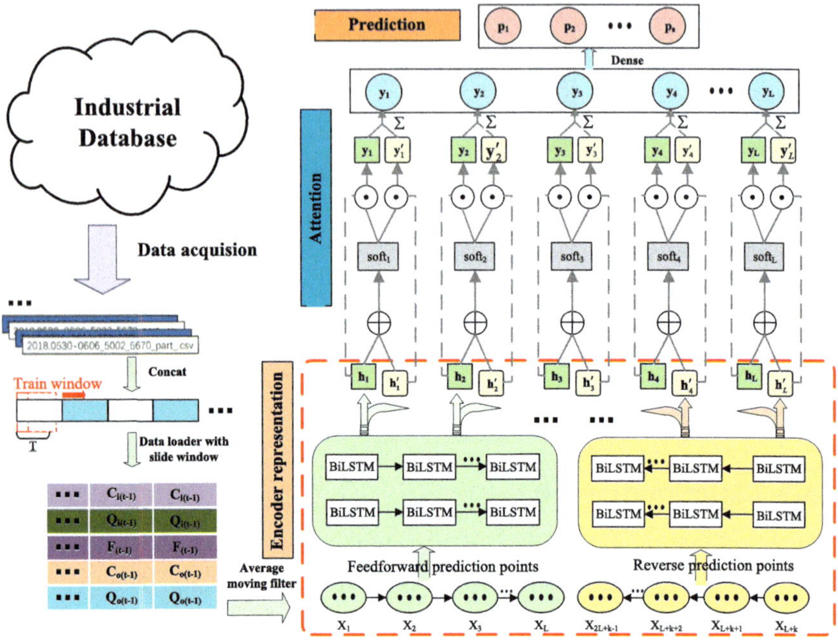

Figure 13.2 Flowchart of the proposed underflow concentration prediction method for the deep cone thicker production.

Suppose the feedforward and reverse input weights $\vec{\mathbf{W}}_z$, $\vec{\mathbf{W}}_i$, $\vec{\mathbf{W}}_f$, $\vec{\mathbf{W}}_o$, $\overleftarrow{\mathbf{W}}_z$, $\overleftarrow{\mathbf{W}}_i$, $\overleftarrow{\mathbf{W}}_f$, $\overleftarrow{\mathbf{W}}_o \in \mathbb{R}^{N\times M}$; recurrent feedforward and reverse weights $\vec{\mathbf{R}}_z, \vec{\mathbf{R}}_i, \vec{\mathbf{R}}_f, \vec{\mathbf{R}}_o, \overleftarrow{\mathbf{R}}_z, \overleftarrow{\mathbf{R}}_i, \overleftarrow{\mathbf{R}}_f, \overleftarrow{\mathbf{R}}_o \in \mathbb{R}^{N\times N}$; output feedforward and reverse weights $\vec{\mathbf{p}}_i, \vec{\mathbf{p}}_f, \vec{\mathbf{p}}_o, \vec{\mathbf{p}}_i, \overleftarrow{\mathbf{p}}_f, \overleftarrow{\mathbf{p}}_o \in \mathbb{R}^N$; and bias weights $\vec{\mathbf{b}}_z, \vec{\mathbf{b}}_i, \vec{\mathbf{b}}_f, \vec{\mathbf{b}}_o, \overleftarrow{\mathbf{b}}_z, \overleftarrow{\mathbf{b}}_i, \overleftarrow{\mathbf{b}}_f, \overleftarrow{\mathbf{b}}_o \in \mathbb{R}^N$.

The piecewise epoch equations $\mathbf{z}$ are expressed as

$$
\begin{aligned}
\overline{\mathbf{z}}^t{}_f &= \vec{\mathbf{W}}_z \mathbf{x}_f{}^t + \vec{\mathbf{R}}_z \mathbf{y}_f{}^{t-1} + \vec{\mathbf{b}}_z, \\
\mathbf{z}_f{}^t &= g(\overline{\mathbf{z}}^t{}_f).
\end{aligned}
\tag{13.2}
$$

The dual forward and reverse input gates $\mathbf{i}$s are

$$
\begin{aligned}
\bar{\mathbf{i}}_f{}^t &= \vec{\mathbf{W}}_i \mathbf{x}_f{}^t + \vec{\mathbf{R}}_i \mathbf{y}_f{}^{t-1} + \vec{\mathbf{p}}_i \odot \vec{\mathbf{c}}^{t-1} + \vec{\mathbf{b}}_i, \\
\mathbf{i}_f{}^t &= \sigma(\bar{\mathbf{i}}_f^t),
\end{aligned}
\tag{13.3}
$$

$$\begin{aligned}\bar{\mathbf{i}}_r^{\,t} &= \overleftarrow{\mathbf{W}}_i\mathbf{x}_r^{\,t} + \overleftarrow{\mathbf{R}}_i\mathbf{y}_r^{\,t-1} + \overleftarrow{\mathbf{p}}_i \odot \overleftarrow{\mathbf{c}}^{\,t-1} + \overleftarrow{\mathbf{b}}_i,\\ \mathbf{i}_r^{\,t} &= \sigma(\overleftarrow{\mathbf{i}}_r).\end{aligned} \tag{13.4}$$

The dual forward and reverse forgot gates **f**s are

$$\begin{aligned}\bar{\mathbf{f}}_f^{\,t} &= \vec{\mathbf{W}}_f\mathbf{x}_f^{\,t} + \vec{\mathbf{R}}_f\mathbf{y}_f^{\,t-1} + \vec{\mathbf{p}}_f \odot \vec{\mathbf{c}}^{\,t-1} + \vec{\mathbf{b}}_f,\\ \mathbf{f}_f^{\,t} &= \sigma(\bar{\mathbf{f}}_f^{\,t}),\end{aligned} \tag{13.5}$$

$$\begin{aligned}\bar{\mathbf{f}}_r^{\,t} &= \overleftarrow{\mathbf{W}}_f\mathbf{x}_r^{\,t} + \overleftarrow{\mathbf{R}}_f\mathbf{y}_r^{\,t-1} + \overleftarrow{\mathbf{p}}_f \odot \overleftarrow{\mathbf{c}}^{\,t-1} + \overleftarrow{\mathbf{b}}_f,\\ \mathbf{f}_r^{\,t} &= \sigma(\bar{\mathbf{f}}_r^{\,t}).\end{aligned} \tag{13.6}$$

Then the forward and reverse cell states can be represented, respectively, by

$$\begin{aligned}\mathbf{c}_f^{\,t} &= \mathbf{z}_f^{\,t} \odot \mathbf{i}_f^{\,t} + \mathbf{c}_f^{\,t-1} \odot \mathbf{f}_f^{\,t},\\ \mathbf{c}_r^{\,t} &= \mathbf{z}_r^{\,t} \odot \mathbf{i}_r^{\,t} + \mathbf{c}_r^{\,t-1} \odot \mathbf{f}_r^{\,t}.\end{aligned} \tag{13.7}$$

The forward and reverse outputs **o**s are

$$\begin{aligned}\bar{\mathbf{o}}_f^{\,t} &= \vec{\mathbf{W}}_o\mathbf{x}_f^{\,t} + \vec{\mathbf{R}}_o\mathbf{y}_f^{\,t-1} + \vec{\mathbf{p}}_o \odot \mathbf{c}_f^{\,t} + \vec{\mathbf{b}}_o,\\ \mathbf{o}_f^{\,t} &= \sigma(\bar{\mathbf{o}}_f^{\,t}),\end{aligned} \tag{13.8}$$

$$\begin{aligned}\bar{\mathbf{o}}_r^{\,t} &= \overleftarrow{\mathbf{W}}_o\mathbf{x}_r^{\,t} + \overleftarrow{\mathbf{R}}_o\mathbf{y}_r^{\,t-1} + \overleftarrow{\mathbf{p}}_o \odot \mathbf{c}_r^{\,t} + \overleftarrow{\mathbf{b}}_o,\\ \mathbf{o}_r^{\,t} &= \sigma(\bar{\mathbf{o}}_r^{\,t}),\end{aligned} \tag{13.9}$$

$$\begin{aligned}h_i^{\,t} &= h(\mathbf{c}_f^{\,t}) \odot \mathbf{o}_f^{\,t},\\ (h')_i^{\,t} &= h(\mathbf{c}_r^{\,t}) \odot \mathbf{o}_r^{\,t},\end{aligned} \tag{13.10}$$

where σ, g, and h are the piecewise nonlinear activation projections, respectively. Since the backward and forward processes are the same in principle but in opposite series order, the equation of the backward process could be derived similarly by replacing $\rightarrow$ with $\leftarrow$. When we finish the encoder representation of the underflow key quality features, the attention mechanism inherited from the BiLSTM output can be given as the following embedding feedforward process:

$$soft_i = \tanh\left(\vec{W}_h h_i + \vec{b}_h\right) + \tanh\left(\overleftarrow{W}_h h'_i + \overleftarrow{b}_h\right), \tag{13.11}$$

$$\alpha_i = \frac{\exp\left(soft_i\right)}{\sum_{t=1}^{N}\exp\left(soft_t\right)}, \quad \sum_{i=1}^{N}\alpha_i = 1, \tag{13.12}$$

$$y = \sum_{i=1}^{N} \alpha_i h_i, \quad y' = \sum_{i=1}^{N} \alpha_i h_i'. \tag{13.13}$$

Then the final output of DualLSTM would be

$$p_j = \sum_{j=1}^{s} W_j \cdot y_i \langle 1, \ldots L \rangle = \sum_{j=1}^{s} \sum_{i=1}^{L} W_j W_y \cdot \langle y_i, y_i' \rangle, \tag{13.14}$$

where W_j and W_y are the weight parameters that connect the dense and attention layers, respectively. Thus the aim of Eqs. (13.2)–(13.14) is to learn the mapping prediction function

$$\begin{aligned} & p\left(y_{t+s} \mid X_1, X_2, \ldots X_L, X_{L+k}, X_{L+k+1}, \ldots, X_{2L+k-1}\right) \\ & = p(y_{t+s} \mid \langle \mathbf{C}_{i(t)}, \mathbf{Q}_{i(t)}, \mathbf{F}_{(t)}, \mathbf{C}_{o(t)}, \mathbf{Q}_{o(t)} \rangle) \\ & = p(y \mid W \cdot \sum_{i=1}^{L} \langle y_i, y_i' \rangle). \end{aligned} \tag{13.15}$$

The backpropagation algorithm with Adam optimizer is employed to train the corresponding parameters in the DualLSTM model.

In this framework, a feedforward and reverse bi-directional stacked long short-term memory is constructed for obtaining a complex temporal feature extraction. Then the attention mechanism is integrated for the robust feature aggregation, and the output for the prediction is obtained from the high-level abstract features.

13.4 Numerical examples

During this period, through training experiments, the influence of training batch size and sliding window length T on prediction accuracy was explored. The corresponding comparison results are shown in Fig. 13.3. On the way, 3000 epochs were trained, the three layers of BiLSTM layer that concatenates the hidden features to the encoder representation, and the hidden layer size is 256. The sliding window length and the batch size are set to 4 and 8, respectively.

The data collected directly by the actual industrial sensor are subject to certain restrictions such as disturbances, random errors, or occasional sudden outliers, which is the original motivation for developing the proposed model. In the proposed DualLSTM framework, this problem can be alleviated with appropriate filter processing. Experimental results show that using a moving average filter of length [20, 40, 20, 10, 20] can significantly improve the prediction accuracy and reduce the root mean square error (about −96%). Fig. 13.3 evaluates the measured evaluation loss indexes comparison for different algorithms with GRU, LSTM, ANN, and Ours, and RMSE loss is also illustrated, see Fig. 13.3. The traditional ANN has the largest loss during the training process. It also further verifies the prediction efficacy of the proposed framework. All the experiments are conducted based on the 95% confidence interval. In Fig. 13.3, compared to the LSTM and GRU, our proposed DualLSTM acquires smoother prediction performance in the training dataset. Fig. 13.3(e) outperforms Fig. 13.3(a)–(d) in the testing dataset performance, which means that our proposed DualLSTM achieves the highest accurate prediction. It compares the DualLSTM's whole performance in the training, validation, and testing datasets. It shows that the proposed DualLSTM gives a more stable and robust prediction (the confidence interval is smaller than that of BiLSTM), which means that underflow concentration prediction is more confidential.

Table 13.1 Evaluation indexes for various competitive prediction methods.

Methods	RMSE	MAE
RNN	0.6712 + 0.0124	0.3488 + 0.0082
GRU	0.4634 + 0.0078	0.2499 + 0.0345
LSTM	0.7533 + 0.0030	0.5119 + 0.0079
SVR	0.8133 + 0.0002	0.6003 + 0.0013
LGB	0.8091 + 0.0092	0.6083 + 0.0118
VAEWGAN	0.7985 + 0.0116	0.5847 + 0.0164
GSTAE	0.7735 + 0.0068	0.5296 + 0.0097
SS-PdeepFM	0.7650 + 0.0036	0.5206 + 0.0073
MSWR-LRCN	0.7571 + 0.0033	0.5235 + 0.0092
LSTM-DeepFM	0.7497 + 0.0026	0.5091 + 0.0071
MPA-RNN	0.7515 + 0.0034	0.2074 + 0.0072
DA-RNN	0.851 + 0.0029	0.2322 + 0.0052
Attention	0.9086 + 0.0021	0.2445 + 0.0031
DSTP-RNN	0.8496 + 0.0051	0.2254 + 0.0042
Suggested DualLSTM	**0.2234 + 0.0021**	**0.1731 + 0.0013**

As shown in Table 13.1, compared to the other competitive methods, DualLSTM shows the best performance with the lowest RMSE and MAE, 0.2234 and 0.1735, respectively. The SVR with a multi-polynomial

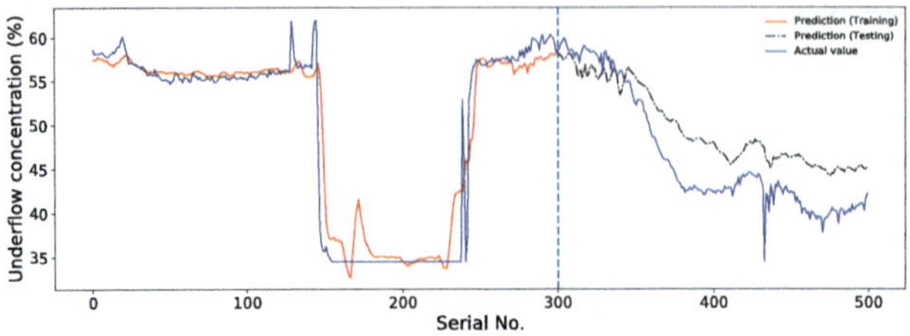

(a)

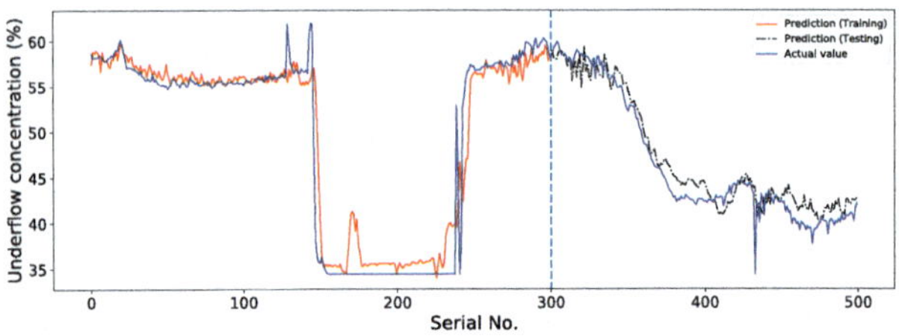

(b)

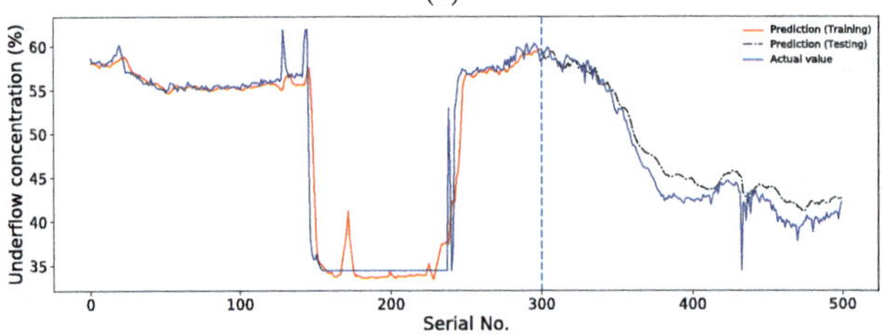

(c)

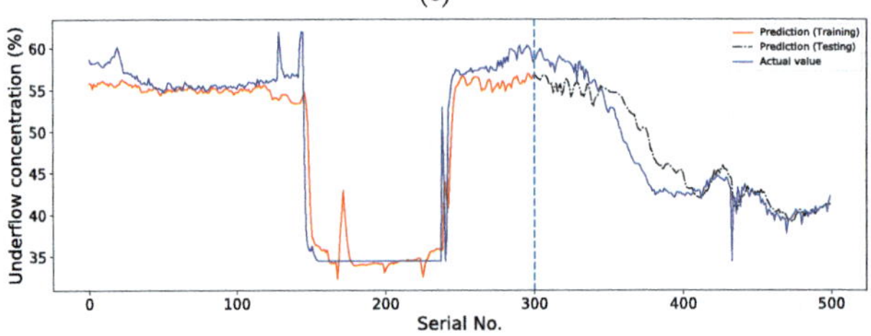

(d)

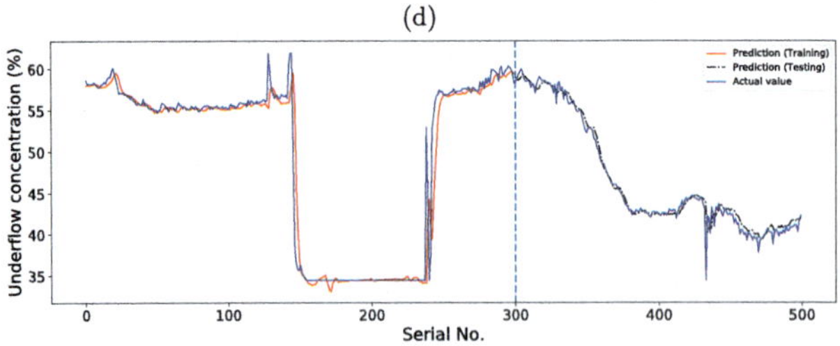

(e)

Figure 13.3 The underflow prediction comparison with competitive algorithms (refer to paper B).

activation function achieves the second-best performance, compared with the basic LSTM and GRU networks. However, if we use other kernels, like linear kernel or Gaussian kernel, then the RMSE and MAE are very high (0.9763 in RMSE, 0.7236 in MAE, respectively). In our industrial case, the performance of the LSTM is lower than that of the GRU network. Compared with the attention additive, the performance was improved by a satisfactory amount, which means that the attention mechanism learns other complex representations with different attention in the underflow concentration process variables. The multi-layer attention in those modules also helped improve the overall prediction performance.

On the other side, the addition of the average moving filter in the industrial case study shows that the prediction accuracy was improved by a large margin (approximately 90%). The different sliding time window was used in our experiment, and the result shows that the best parameters are [20, 40, 20, 10, 20]. In DualLSTM the front of the underflow concentration instant is considered to train the whole model with the attention mechanism. The experiment results show that we have leveraged the proposed prediction model in a state-of-the-art fashion. The other prediction model is also compared in our experiments; the LSTM's performance outperforms the RNN-based model, and the specific RMSE and MAE are 0.7533, and 0.5119, respectively. Our DualLSTM outperforms the other prediction methods because the hidden information from the hidden variables is fully captured by the proposed average moving filter, encoder, attention, and finally softmax transmission without loss. Besides, the front underflow concentration and reverse input variables are mutually and jointly to be used to train the whole and achieve superior performance and robustness. A remarkable performance has been achieved in this proposed architecture.

13.5 Conclusions

This chapter provides a high-efficiency key-quality underflow prediction model for the deep cone thickener system to address the challenges of underflow prediction in industrial CTS plants. In the industrial case, compared with other competitive algorithms, the proposed DualLSTM proposed a new architecture with feedforward and inverse dual BiLSTM, and the final MAE and RMSE are 0.2234 and 0.1731, respectively. The proposed algorithms additionally take into account time-invariant elements and improved the prediction performance by a large margin. DCT's underflow concentration prediction assesses the presented algorithms' industrial feasibility and effectiveness. The DualLSTM can be extended to implement some other similar process industrial domains such as nonferrous, process manufacturing, and chemical production. For further research, we will continue to investigate this model and apply it to optimal control and establish a comprehensive intelligent integrated platform to promote the development of the smart mining and paste-filling industry.

References

[1] Yongxiang Lei, Hamid Reza Karimi, Underflow concentration prediction based on improved dual bidirectional LSTM for hierarchical cone thickener system, The International Journal of Advanced Manufacturing Technology (2023) 1–12.

[2] Yongxiang Lei, Intelligent optimal framework for the industrial mining plant-wide prediction control, IEEE Transactions on Instrumentation and Measurement (2023).

[3] Hong Li, Ai-xiang Wu, Hong-Jiang Wang, Hui Chen, Liu-Hua Yang, Changes in underflow solid fraction and yield stress in paste thickeners by circulation, International Journal of Minerals, Metallurgy, and Materials 28 (3) (2021) 349–357.

[4] Dong Xiao, Hongfei Xie, Longqiang Jiang, Ba Tuan Le, Jichun Wang, Chongmin Liu, Hongzong Li, Research on a method for predicting the underflow concentration of a thickener based on the hybrid model, Engineering Applications of Computational Fluid Mechanics 14 (2020) 13–26, https://doi.org/10.1080/19942060.2019.1658228.

[5] Zhaolin Yuan, Jinlong Hu, Di Wu, Xiaojuan Ban, A dual-attention recurrent neural network method for deep cone thickener underflow concentration prediction, Sensors 20 (5) (2020) 1260.

[6] Yongxiang Lei, Hamid Reza Karimi, Xiaofang Chen, A novel self-supervised deep LSTM network for industrial temperature prediction in aluminum processes application, Neurocomputing 502 (2022) 177–185.

[7] Ping Wang, Yichao Yin, Xiaogang Deng, Yingchun Bo, Weiming Shao, Semi-supervised echo state network with temporal–spatial graph regularization for dynamic soft sensor modeling of industrial processes, ISA Transactions 130 (2022) 306–315.

[8] Ronil Barua, Anil K. Sharma, Dynamic black litterman portfolios with views derived via CNN-BiLSTM predictions, Finance Research Letters 49 (2022) 103111.

[9] Shuo Sun, Junzhong Sun, Zongliang Wang, Zhiyong Zhou, Wei Cai, Prediction of battery SOH by CNN-BiLSTM network fused with attention mechanism, Energies 15 (12) (2022) 4428.

[10] Wenshuai Lin, Bin Zhang, Hongyi Li, Renquan Lu, Multi-step prediction of photovoltaic power based on two-stage decomposition and BILSTM, Neurocomputing 504 (2022) 56–67.

[11] Tao Liang, Chunjie Chai, Hexu Sun, Jianxin Tan, Wind speed prediction based on multi-variable Capsnet-BiLSTM-MOHHO for WPCCC, Energy 250 (2022) 123761.

[12] Hui Hwang Goh, Ronghui He, Dongdong Zhang, Hui Liu, Wei Dai, Chee Shen Lim, Tonni Agustiono Kurniawan, Kenneth Tze Kin Teo, Kai Chen Goh, A multimodal approach to chaotic renewable energy prediction using meteorological and historical information, Applied Soft Computing 118 (2022) 108487.

[13] Lei Chen, Weiwen Zhang, Haiming Ye, Accurate workload prediction for edge data centers: Savitzky–Golay filter, CNN and BiLSTM with attention mechanism, Applied Intelligence 52 (11) (2022) 13027–13042.

[14] Chu Zhang, Huixin Ma, Lei Hua, Wei Sun, Muhammad Shahzad Nazir, Tian Peng, An evolutionary deep learning model based on TVFEMD, improved sine cosine algorithm, CNN and BiLSTM for wind speed prediction, Energy 254 (2022) 124250.
[15] Kaixiong Hu, Yanghui Wang, Weidong Li, Lihui Wang, CNN-BiLSTM enabled prediction on Molten pool width for thin-walled part fabrication using laser directed energy deposition, Journal of Manufacturing Processes 78 (2022) 32–45.
[16] Wenbai Chen, Weizhao Chen, Huixiang Liu, Yiqun Wang, Chunli Bi, Yu Gu, A RUL prediction method of small sample equipment based on DCNN-BiLSTM and domain adaptation, Mathematics 10 (7) (2022) 1022.
[17] L. Shuaiyi, Kai Wang, Liren Zhang, Bailing Wang, Global-local integration for GNN-based anomalous device state detection in industrial control systems, Expert Systems with Applications 209 (2022) 118345.
[18] Yao Qin, Dongjin Song, Haifeng Chen, Wei Cheng, Guofei Jiang, Garrison Cottrell, A dual-stage attention-based recurrent neural network for time series prediction, arXiv preprint, arXiv:1704.02971, 2017.
[19] Sepp Hochreiter, Jürgen Schmidhuber, Long short-term memory, Neural Computation 9 (8) (1997) 1735–1780.
[20] Chaoshun Li, Geng Tang, Xiaoming Xue, Adnan Saeed, Xin Hu, Short-term wind speed interval prediction based on ensemble GRU model, IEEE Transactions on Sustainable Energy 11 (3) (2019) 1370–1380.
[21] Max Jaderberg, Karen Simonyan, Andrew Zisserman, et al., Spatial transformer networks, Advances in Neural Information Processing Systems 28 (2015).
[22] Chenliang Liu, Kai Wang, Yalin Wang, Xiaofeng Yuan, Learning deep multimanifold structure feature representation for quality prediction with an industrial application, IEEE Transactions on Industrial Informatics 18 (9) (2021) 5849–5858.
[23] Yuchen Jiang, Shen Yin, Recent advances in key-performance-indicator oriented prognosis and diagnosis with a MATLAB toolbox: DB-KIT, IEEE Transactions on Industrial Informatics 15 (5) (2018) 2849–2858.
[24] Xiuli Zhu, Seshu Kumar Damarla, Kuangrong Hao, Biao Huang, Hongtian Chen, Yicun Hua, ConvLSTM and self-attention aided canonical correlation analysis for multi-output soft sensor modeling, IEEE Transactions on Instrumentation and Measurement (2023).
[25] Biao Zhang, Deyi Xiong, Jinsong Su, Neural machine translation with deep attention, IEEE Transactions on Pattern Analysis and Machine Intelligence 42 (1) (2018) 154–163.
[26] Jin Chen, Xingchen Li, Ye Xiao, Hao Chen, Yong Zhao, FRA-LSTM: a vessel trajectory prediction method based on fusion of the forward and reverse sub-network, arXiv preprint, arXiv:2201.07606, 2022.
[27] J. Geng, C. Yang, Y. Li, et al., MPA-RNN: a novel attention-based recurrent neural networks for total nitrogen prediction, IEEE Transactions on Industrial Informatics 18 (10) (2022) 6516–6525.
[28] Klaus Greff, Rupesh K. Srivastava, Jan Koutník, Bas R. Steunebrink, Jürgen Schmidhuber, LSTM: a search space odyssey, IEEE Transactions on Neural Networks and Learning Systems 28 (10) (2016) 2222–2232.
[29] Xiaofeng Yuan, Lin Li, Yuri A.W. Shardt, Yalin Wang, Chunhua Yang, Deep learning with spatiotemporal attention-based LSTM for industrial soft sensor model development, IEEE Transactions on Industrial Electronics 68 (5) (2020) 4404–4414.
[30] Chenliang Liu, Kai Wang, Yalin Wang, Shengli Xie, Chunhua Yang, Deep nonlinear dynamic feature extraction for quality prediction based on spatiotemporal neighborhood preserving SAE, IEEE Transactions on Instrumentation and Measurement 70 (2021) 1–10.

[31] Yongxiang Lei, Hamid Reza Karimi, Lihui Cen, Xiaofang Chen, Yongfang Xie, Processes soft modeling based on stacked autoencoders and wavelet extreme learning machine for aluminum plant-wide application, Control Engineering Practice 108 (2021) 104706.
[32] Xiaofeng Yuan, Shuaibin Qi, Yalin Wang, Haibing Xia, A dynamic CNN for nonlinear dynamic feature learning in soft sensor modeling of industrial process data, Control Engineering Practice 104 (2020) 104614.
[33] Wei Xie, Jiesheng Wang, Cheng Xing, Shasha Guo, Mengwei Guo, Lingfeng Zhu, Variational autoencoder bidirectional long and short-term memory neural network soft-sensor model based on batch training strategy, IEEE Transactions on Industrial Informatics 17 (8) (2020) 5325–5334.

CHAPTER 14

A deeply efficient long short-term memory framework for underflow concentration prediction: DE-LSTM

14.1 Introduction

Paste filling in the mining industry has received high attention in the development of power generation. For sustainable environment protection and the carbon neutral and zero emission target, industrial mining is making a revolution toward smart, intelligent, and clean underground mining [1–5]. Deep cone thickener (DCT) is a critical device to produce underground pasting for mining security. The main principle of the device is that the low-concentration slurry crude from the stope was fed into the DCT, and the gravity of these particles is then dissolved to a larger lump under the effect of flocculant and concentrated at the bottom of the thickener. The high-quality underflow concentration is then produced, and the high layer of clean water is recycled by the overflow pipe.

Stable underflow concentration is a critical metric, which plays a pivotal role in assessing the performance, security, and stability of industrial production processes. Several quality factors, including the volume of underflow, feed flow, flocculant dosage, and mass of flocculant, significantly influence the stability of underflow concentration during the production process. Therefore ensuring high performance and robustness in underflow prediction and control is crucial for efficient CTS production. However, conventional detection techniques fall short in enabling online prediction, while their high computational costs and susceptibility to sampling oscillation are added to this challenge. By analyzing the production mechanism it is evident that production variables have a direct correlation with the output underflow concentration, highlighting the need for a cost-effective data-centric approach for underflow concentration prediction [6].

For subsurface paste filling, a deep cone thickener is essential. CTS is a key method for achieving a consistent concentration for underground

Learning-Based Soft Sensing and Predictions for Process Industries
https://doi.org/10.1016/B978-0-44-336759-5.00023-8

mining fills. Pipe-blocking mishaps may occur during the thickening process if the subterranean concentration is too high. On the other hand, a low subterranean concentration reduces the quality of the entire backfilled paste and affects the overall safety of the mining operation. Hence the development of a data-driven model for predicting underflow concentration in the CTS system is of utmost importance. The whole mining paste filling operates on a continuous hierarchical concept [7]. The CTS was fed with the crude unstable low-concentration slurry flow (almost 20–30%), which was combined with a flocculant to speed up the sinking rate. A mud bed can collect the dissolved particles. The suitable concentration and volume feed flow are created at the bottom of the CTS. The top clean water from the overflow pipe has also been recycled for future use. The primary control of CTS provides a consistent and precise underflow concentration [8]. The concentration of underflow is a key metric for assessing the effectiveness and efficiency of the industrial underground pasting process. It is necessary to identify changes in the different variables due to the inner link for the production quality variable in the deep cone thickening process. Because the response has a long time lag, various traditional methods have also been used as the fundamental unit. This time-series architecture may be used to acquire some prior knowledge and historical information, which can subsequently be used to anticipate the underflow concentration.

Recently, deep learning has become a hot topic in the modeling, optimization, control, and prediction of industrial systems. For instance, Lei et al. proposed a semi-supervised method to tackle the superheat degree identification in the aluminum electrolysis reduction cell, which has a corresponding relation with our industrial case. Yuan et al. investigated a long short-term memory with attention mechanism with the time sequence soft sensor tasks. Some other methods, such as self-supervised learning, have also achieved large attention in the industrial process domain [9–14]. Although recurrent neural networks can retain historical information using memory and nonlinear functions, they suffer from low accuracy, gradient varnishing, and exploration issues due to their long dependency on historical features [15–19]. Long short-term memory (LSTM) is an enhanced version of RNN that overcomes these limitations by introducing a new memory unit to store historical information [20,21]. This makes LSTM ideal for practical applications, as it covers the input gate, cell gate, forget gate, and output gate, retaining only relevant information. In practical applications the attention network is used for the hierarchical industrial process. Other LSTM-based applications with attention to the mechanism are given in [22–25]; however, the spatial information is largely ignored. Other recent variants of long short-term memory are presented in [26–30].

In particular, several time-sequence models can capture valuable historical information that should be effectively used. The prediction of underflow concentration is a crucial challenge in the field of industrial cone thickeners. Recently, some works based on the underflow concentration are reported. For example, Yuan et al. proposed a dual-attention recurrent neural network method for the deep cone thickener; their model can capture the spatial information with high-dimensional series. To our best knowledge, some underflow concentration detection methods have been proposed [31–34]; however, they only consider the static feature prediction, and the adaptive and dynamic mechanisms for online learning are largely ignored, so the time-series long prediction for the CTS system still needs to be further investigated.

Some conventional time-series prediction models are widely used in the industrial analysis, such as the principal component analysis (PCA), autoregressive integrated moving average (ARIMA), recurrent neural network, and gated recurrent networks [35–39]. All these architectures have been widely used in industrial fields [40–42]. In the deep cone thickening system, some basic hypotheses are proposed to be considered: (1) The coupled multi-variables are linked with the output underflow concentration variables, which means that we can build a data-driven model to learn the complex nonlinear mapping between those variables; (2) The large delay in the industrial case gives the insight that we should propose a time sequence to learn the time series information; (3) The industrial database collected from all the sensors has outliers and noise, which greatly reduces the prediction performance in the applicable CTS. To tackle these problems, we propose a new LSTM method. First, the original data from the industrial database are preprocessed by an average moving filter. Second, the encoder transmission channel was leveraged to gain the feature representation of the underflow concentration prediction information. Finally, the attention mechanism with a fully connected layer is used for the final underflow prediction in the CTS system. The main contributions of this chapter can be highlighted as follows:

1. A novel DE-LSTM model is proposed to deal with the long-time prediction with robustness and high accuracy for the large time delay CTS pasting system.

2. Compared to the traditional prediction methods such as SVM, LSTM, GRU, and XGBOOST, the performance of the proposed DE-LSTM provides a more accurate and robust result.
3. Stacked long short-term memory with attention mechanism is employed for the long prediction without information loss during the learning process.
4. The industrial application case is proposed to evaluate the high performance of the proposed DE-LSTM method. The experiments verify that the prediction MSE has decreased by a large margin with the presented framework.

14.2 Problem statement

In the underground paste-filling process, the underflow concentration is the top index, which can sustain the stable production of the paste, avoid the security hazard, and improve the quality of the paste-filling. The detailed structure of the deep cone thickener is presented in Fig. 14.1, and all the processes in the pasting-filling process are included. However, the traditional method for prediction accuracy is limited by the expensive physical device, the large time delay, and the difficulty of online detection. In the long term, the flocculant should react with the feed flow particle for a long time, so the change of the underflow concentration is a long time interval. To the best of our knowledge, the existing production of intelligent DCT systems is scarce, and the tendency of the deep cone thickener is a smart mining industry and paste filling.

Based on the above analysis, the problem can be specified as follows: Given multi-dimension time input variables $(x_1, x_2, \ldots, x_t)$ and temporal underflow concentration $(y_1, y_2, \ldots, y_t)$, we need to learn a distribution when we observe all the multi-dimensional massive input variables. The specific expression is

$$p\left(y_{t+L}, y_{t+L-1}, \ldots, y_{t+1} \,\middle|\, y_1, y_2, \ldots, y_t, x_1, x_2, \ldots, x_t\right),$$

where T is the sliding moving window size for the underflow future prediction. The aim is to achieve a model with the nonlinear sequential mapping function expressed as follows:

$$\max p\left(y_{t+L}, y_{t+L-1}, \ldots, y_{t+1} \,\middle|\, y_1, y_2, \ldots, y_t, x_1, x_2, \ldots, x_t\right) = f(\cdot), \quad (14.1)$$

where the $\cdot$ represents the multi-dimension variables. The main objective is to obtain high performance in the long interval prediction accuracy and robustness.

14.3 Main framework

Before the construction of an excellent DE-LSTM model, the preprocessing of the collected data is necessary. The original data from the massive sensors have the oscillation points, outlier, and missing data, which greatly influence the accuracy of the proposed DE-LSTM model. Before considering the sliding window to train the data, the standardization and normalization of the training samples and the temporal output are used for improving the quality of the data. The feed flow concentration, flocculant rate, and feed amount are the initial input variables.

The second target is the encoder representation of the underflow concentration prediction in the CTS system. The historical information should be preserved in the long temporal term. The LSTM unit is built to learn the temporal information contained in the training samples. The specific representation inherited from the front layer is

$$\max p\left(y_{t+L}, y_{t+L-1}, \ldots, y_{t+1}/y_1, y_2, \ldots, y_t, x_1, x_2, \ldots, x_t\right) = LSTM\left(\cdot\right). \tag{14.2}$$

The features from the average moving filter are sent to the deep encoder network to obtain an efficient representation of the underflow concentration. Given the time series of $(x_1, x_2, \ldots, x_L)$, the hidden representation from the multiple stacked LSTM can be represented as

$$h_i = LSTM\left(x_1, x_2, \ldots, x_L\right). \tag{14.3}$$

The details of this equation can be extended as

$$\tilde{\mathbf{i}}_t = \sigma(\mathbf{W}_{xi}\mathbf{x}_t + \mathbf{W}_{hi}\tilde{\mathbf{h}}_{t-1} + \mathbf{W}_{ci}\tilde{\mathbf{c}}_{t-1} + \mathbf{b}_i), \tag{14.4}$$

$$\tilde{\mathbf{f}}_t = \sigma(\mathbf{W}_{xf}\mathbf{x}_t + \mathbf{W}_{hf}\tilde{\mathbf{h}}_{t-1} + \mathbf{W}_{cf}\tilde{\mathbf{c}}_{t-1} + \mathbf{b}_f), \tag{14.5}$$

$$\tilde{\mathbf{c}}_t = \tilde{\mathbf{f}}_t \odot \tilde{\mathbf{c}}_{t-1} + \tilde{\mathbf{i}}_t \odot \tanh(\mathbf{W}_{xc}\mathbf{x}_t + \mathbf{W}_{hc}\tilde{\mathbf{h}}_{t-1} + \mathbf{b}_c), \tag{14.6}$$

$$\tilde{\mathbf{o}}_t = \sigma(\mathbf{W}_{xo}\mathbf{x}_t + \mathbf{W}_{ho}\tilde{\mathbf{h}}_{t-1} + \mathbf{W}_{co}\tilde{\mathbf{c}}_t + \mathbf{b}_o), \tag{14.7}$$

$$\tilde{\mathbf{h}}_t = \tilde{\mathbf{o}}_t \odot \tanh\left(\tilde{\mathbf{c}}_t\right). \tag{14.8}$$

The encoder representation extracts complex features and learns historical temporal information from the underflow concentration variables. The attention mechanism process from the encoder layer is expressed as

$$M = \sum_{i=1}^{N} \tanh\left(h_i\right), \tag{14.9}$$

and the transmission of the front layer is written as

$$\alpha = \frac{w_a^T M}{\sum_i \left(w_a^T M\right)_i}. \tag{14.10}$$

Then the final output of the DE-LSTM is

$$\hat{y} = \sigma\left(\sum_{i=1}^{N} \alpha_i h_i\right). \tag{14.11}$$

The next step is to train the proposed network and gain the weights of the network. The loss function is defined as the least square algorithm, which is specified as follows:

$$Loss = \sum_{i=1}^{N} \sum_{t=1}^{T} \left(\hat{y}_t^i - y_t^i\right)^2, \tag{14.12}$$

where N is the total number of samples, and T is the size of the sliding moving window. The hidden parameters of the LSTM and the attention learnable parameters can be fine-tuned by the Adam optimizer, and he backpropagation is used for the proposed DE-LSTM training.

The flowchart of the proposed DE-LSTM is shown in Fig. 14.1. The procedure is that the original data collected from the industrial CTS database are reprocessed by the sliding window and split with the training datasets. The training data sets with time series T are then transmitted to the average moving filter. The final output of the T steps from DE-LSTM is the prediction of the underflow concentration. Specifically, the dynamic recurrent update mechanism is used in the whole model. The prediction results with the front weights are fine-tuned by a time interval of 1 hour.

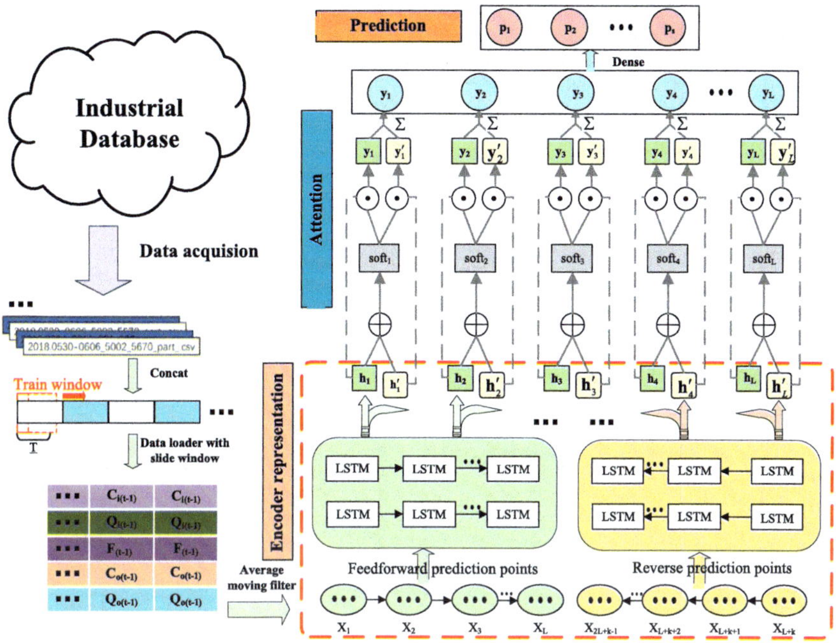

Figure 14.1 The whole framework of the proposed DE-LSTM for the underflow concentration prediction.

14.4 Numerical experiments

To evaluate the performance of the proposed method, we studied two cases. The proposed DE-LSTM is evaluated by the industrial application of long-time underflow prediction for the CTS system. Some public methods such as support vector regression (SVR), neural networks (NNs), deep belief networks (DBNs), and the temporal recurrent network such as recurrent neural network (RNN), XGBOOST, long short-term memory (LSTM), and gated recurrent unit (GRU) and DARNN are also compared in the experiments. During this period, through training experiments, the influence of training batch size and sliding window length T on prediction accuracy was explored. The results are shown in Fig. 14.2. In the process, 3000 epochs were trained, the three layers of BiLSTM layer that concatenate the hidden features to the encoder representation, and the hidden layer size is 256. The different batch size is used to testify the approximate size to obtain the state-of-the-art prediction performance. In the study of the concentration of the underflow, the batch size is set to 10, 15, 20, and 25, and the corresponding results are reported in Fig. 14.2. The

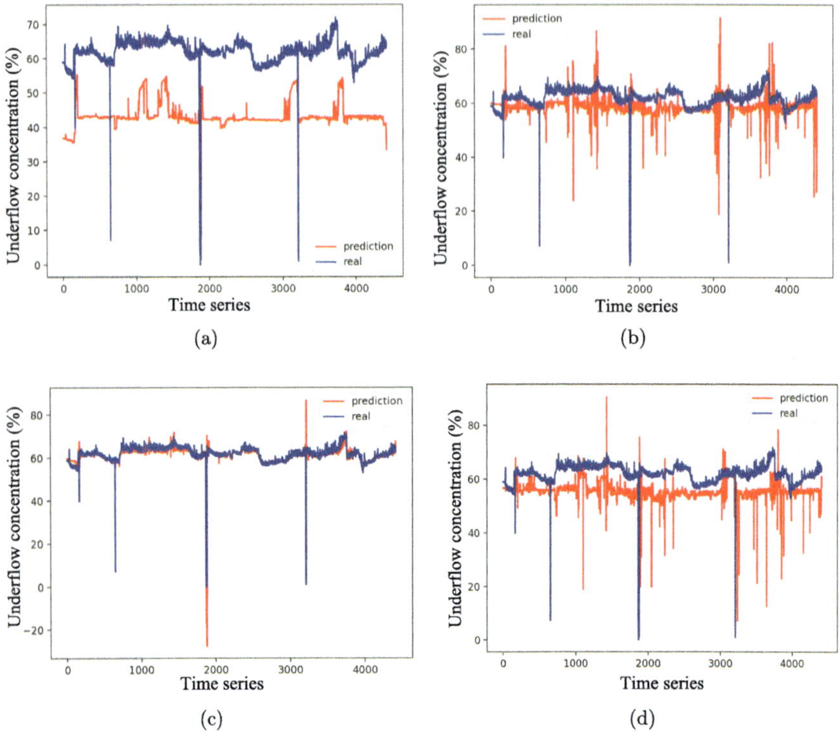

Figure 14.2 The proposed DE-LSTM underflow prediction compared to algorithms (a) SVR; (b) BiLSTM; (c) GRU; (d) LSTM.

experimental results demonstrate that under the batch size constraint equal to 15, the algorithm achieves the best prediction performance. However, the other three comparisons cause unsatisfactory accuracy, especially with a batch size equal to 25. The prediction mean square error is large, and the prediction curve cannot follow the real underflow concentration timidly, which means that the algorithms need to be further fine-tuned.

On the other hand, the existing state-of-the-art competitive algorithms such as the support vector regression, bidirectional long short-term memory, gate recurrent unit, and long short-term memory are used as the benchmarks to verify the performance of the proposed algorithm. Fig. 14.2 gives the full intuitive description of these results. Fig. 14.2(a) is the underflow concentration prediction with support vector regression, the prediction value (marked in red) varies from the real value (marked with blue), whereas the gate recurrent unit's prediction performance

(Fig. 14.2(b)) is better than that of long short-term memory (Fig. 14.2(d)) and bidirectional long-short time memory (Fig. 14.2(b)). The data collected directly by the actual industrial sensor have disturbances, errors, and occasional sudden outliers, which is the original attribute of the device production. In the proposed DE-LSTM framework, this problem can be alleviated with appropriate filtering. Experimental results show that using a moving average filter of length [20, 40, 20, 10, 20] can greatly improve the prediction accuracy and reduce the root mean square error (about −96%).

Compared to the other competitive methods, DE-LSTM shows the best performance with the lowest RMSE and MAE, 0.2234, and 0.1735, respectively. The SVR with a multi-polynomial activation function also achieves a second performance, compared with the basic LSTM and GRU network. However, if we use other kernels, like linear kernel or Gaussian kernel, then the RMSE and MAE are very high (0.9763 in RMSE, 0.7236 in MAE). In our industrial case, the performance of the LSTM is lower than that of the GRU network. Compared with the attention mechanism additive, the performance improved a satisfactory amount, which means that the attention mechanism learns other complex representations and with the different attention in the underflow concentration process variables. The multi-layer attention in those modules also helped improve the whole prediction performance.

On the other side, the addition of the average moving filter in the industrial shows that the prediction accuracy is improved by a large margin ($\approx$ 90%). The different sliding time window was used in our experiment and the result shows that the best parameters are [20, 40, 20, 10, 20]. In DE-LSTM, the front of underflow concentration instant is considered to train the whole model with the attention mechanism. The experiment results show that we have leveraged the proposed prediction model in a state-of-the-art situation. The other prediction model is competitively compared in our experiments, the LSTM performance outperforms that of the RNN-based model, and the specific RMSE and MAE are 0.7533 and 0.5119, respectively. Our DE-LSTM outperforms the other prediction methods because the hidden information from the hidden variables is fully captured by the proposed average moving filter, encoder, attention, and finally softmax transmission without loss. Besides, the front underflow concentration and reverse input variables are mutually and jointly to be used to train the whole and achieve superior performance and robustness. A remarkable performance has been implemented in this proposed architecture.

14.5 Conclusions

In this chapter, we proposed a new data-centric model for the long-time underflow concentration prediction method in a deep cone thickener system, DE-LSTM. DE-LSTM learns the historical and spatial information without information loss during the transmission of underflow concentration features for the paste-filling process. In our practical case, the use of an average moving filter significantly enhances the prediction performance of the model. The DE-LSTM outperforms other state-of-the-art methods with the lowest RMSE and MAE in underflow concentration prediction while maintaining high robustness. Furthermore, the proposed DE-LSTM can be effortlessly extended to improve its applicability in other industrial domains and processes. For future consideration, we will consider the proposed DE-LSTM as the core bench for the intelligent underground mining system and optimal control platform, which will promote the whole development of the mining industry.

References

[1] E. Chen, H. Cao, H. Li, et al., A big data mining approach for environmental emissions prediction of die casting process, The International Journal of Advanced Manufacturing Technology 114 (11–12) (2021) 3779–3791.

[2] X. Gao, Y. Guo, D.A. Hanson, et al., Thermal error prediction of ball screws based on PSO-LSTM, The International Journal of Advanced Manufacturing Technology 116 (5–6) (2021) 1721–1735.

[3] Yongxiang Lei, Xiaofang Chen, Ming Min, et al., A semi-supervised Laplacian extreme learning machine and feature fusion with CNN for industrial superheat identification, Neurocomputing 381 (2020) 186–195.

[4] Yongxiang Lei, Hamid Reza Karimi, Lihui Cen, Xiaofang Chen, Yongfang Xie, Processes soft modeling based on stacked autoencoders and wavelet extreme learning machine for aluminum plant-wide application, Control Engineering Practice 108 (2021) 104706.

[5] W. Huan, L. Ting, C. Yuning, et al., Underflow concentration prediction model of deep-cone thickener based on data-driven, Journal of China Universities of Posts and Telecommunications 26 (6) (2019) 63.

[6] C.K. Tan, R. Setiawan, J. Bao, et al., Studies on parameter estimation and model predictive control of paste thickeners, Journal of Process Control 28 (2015) 1–8.

[7] C. Fang, D. He, K. Li, et al., Image-based thickener mud layer height prediction with attention mechanism-based CNN, ISA Transactions 128 (2022) 677–689.

[8] I. Takacs, G.G. Patry, D. Nolasco, A dynamic model of the clarification-thickening process, Water Research 25 (10) (1991) 1263–1271.

[9] Dong Xiao, Hongfei Xie, Longqiang Jiang, Ba Tuan Le, Jichun Wang, Chongmin Liu, Hongzong Li, Research on a method for predicting the underflow concentration of a thickener based on the hybrid model, Engineering Applications of Computational Fluid Mechanics 14 (2020) 13–26, https://doi.org/10.1080/19942060.2019.1658228.

[10] Zhaolin Yuan, Jinlong Hu, Di Wu, Xiaojuan Ban, A dual-attention recurrent neural network method for deep cone thickener underflow concentration prediction, Sensors 20 (5) (2020) 1260.
[11] Yao Qin, Dongjin Song, Haifeng Chen, Wei Cheng, Guofei Jiang, Garrison Cottrell, A dual-stage attention-based recurrent neural network for time series prediction, arXiv preprint, arXiv:1704.02971, 2017.
[12] Sepp Hochreiter, Jürgen Schmidhuber, Long short-term memory, Neural Computation 9 (8) (1997) 1735–1780.
[13] Chaoshun Li, Geng Tang, Xiaoming Xue, Adnan Saeed, Xin Hu, Short-term wind speed interval prediction based on ensemble GRU model, IEEE Transactions on Sustainable Energy 11 (3) (2019) 1370–1380.
[14] Max Jaderberg, Karen Simonyan, Andrew Zisserman, et al., Spatial transformer networks, Advances in Neural Information Processing Systems 28 (2015).
[15] Biao Zhang, Deyi Xiong, Jinsong Su, Neural machine translation with deep attention, IEEE Transactions on Pattern Analysis and Machine Intelligence 42 (1) (2018) 154–163.
[16] Xiaofeng Yuan, Lin Li, Yuri A.W. Shardt, Yalin Wang, Chunhua Yang, Deep learning with spatiotemporal attention-based LSTM for industrial soft sensor model development, IEEE Transactions on Industrial Electronics 68 (5) (2020) 4404–4414.
[17] Jin Chen, Xingchen Li, Ye Xiao, Hao Chen, Yong Zhao, FRA-LSTM: a vessel trajectory prediction method based on fusion of the forward and reverse sub-network, arXiv preprint, arXiv:2201.07606, 2022.
[18] J. Geng, C. Yang, Y. Li, et al., MPA-RNN: a novel attention-based recurrent neural networks for total nitrogen prediction, IEEE Transactions on Industrial Informatics 18 (10) (2022) 6516–6525.
[19] Klaus Greff, Rupesh K. Srivastava, Jan Koutník, Bas R. Steunebrink, Jürgen Schmidhuber, LSTM: a search space odyssey, IEEE Transactions on Neural Networks and Learning Systems 28 (10) (2016) 2222–2232.
[20] Xiaofeng Yuan, Lin Li, Yalin Wang, Nonlinear dynamic soft sensor modeling with supervised long short-term memory network, IEEE Transactions on Industrial Informatics 16 (5) (2019) 3168–3176.
[21] S. Xiang, Y. Qin, C. Zhu, et al., LSTM networks based on attention ordered neurons for gear remaining life prediction, ISA Transactions 106 (2020) 343–354.
[22] K. Bandara, C. Bergmeir, H. Hewamalage, LSTM-MSNet: leveraging forecasts on sets of related time series with multiple seasonal patterns, IEEE Transactions on Neural Networks and Learning Systems 32 (4) (2020) 1586–1599.
[23] X. Xu, M. Yoneda, Multitask air-quality prediction based on LSTM-autoencoder model, IEEE Transactions on Cybernetics 51 (5) (2019) 2577–2586.
[24] S. Siami-Namini, N. Tavakoli, A.S. Namin, A comparative analysis of forecasting financial time series using ARIMA, LSTM, and BiLSTM, arXiv preprint, arXiv:1911.09512, 2019.
[25] J. Chen, J. Zhang, X. Xu, et al., E-LSTM-D: a deep learning framework for dynamic network link prediction, IEEE Transactions on Systems, Man and Cybernetics: Systems 51 (6) (2019) 3699–3712.
[26] G.T.S. Ho, H.C.W. Lau, C.K.M. Lee, et al., An intelligent production workflow mining system for continual quality enhancement, The International Journal of Advanced Manufacturing Technology 28 (2006) 744–753, Volume corrected based on context.
[27] T. Peng, C. Zhang, J. Zhou, et al., An integrated framework of bi-directional long-short term memory (BiLSTM) based on sine cosine algorithm for hourly solar radiation forecasting, Energy 221 (2021) 119887.
[28] Y. Li, Z. Zhu, D. Kong, et al., EA-LSTM: evolutionary attention-based LSTM for time series prediction, Knowledge-Based Systems 181 (2019) 104785.

[29] Y. Bin, Y. Yang, F. Shen, et al., Describing video with attention-based bidirectional lstm, IEEE Transactions on Cybernetics 49 (7) (2018) 2631–2641.
[30] Y. Wang, M. Huang, X. Zhu, et al., Attention-based LSTM for aspect-level sentiment classification, in: Proceedings of the 2016 Conference on Empirical Methods in Natural Language Processing, 2016, pp. 606–615.
[31] X. Ran, Z. Shan, Y. Fang, et al., An LSTM-based method with attention mechanism for travel time prediction, Sensors 19 (4) (2019) 861.
[32] S.P. Usher, P.J. Scales, Steady state thickener modelling from the compressive yield stress and hindered settling function, Chemical Engineering Journal 111 (2–3) (2005) 253–261.
[33] B. Koopman, K. Cadec, Prediction of thickening capacity using diluted sludge volume index, Water Research 17 (10) (1983) 1427–1431.
[34] M. Sayah, D. Guebil, Z. Al Masry, et al., Robustness testing framework for RUL prediction Deep LSTM networks, ISA Transactions 113 (2021) 28–38.
[35] L.R. Castillo, R.A. Medronho, A simple procedure for design and performance prediction of Bradley and Rietema hydrocyclones, Minerals Engineering 13 (2) (2000) 183–191.
[36] X. Wang, G. Zhang, J. Zhao, et al., Underflow concentration prediction and external structure parameter optimization of deep cone thickener, Journal of Chongqing University. English Edition 38 (3) (2015) 125–132, Volume and pages inferred from standard format.
[37] Y. Yang, Q. Xiong, C. Wu, et al., A study on water quality prediction by a hybrid CNN-LSTM model with attention mechanism, Environmental Science and Pollution Research 28 (39) (2021) 55129–55139.
[38] T. Zhang, X.Q. Zheng, M.X. Liu, Multiscale attention-based LSTM for ship motion prediction, Ocean Engineering 230 (2021) 109066.
[39] Z. Liu, L. Li, X. Fang, et al., Hard-rock tunnel lithology prediction with TBM construction big data using a global-attention-mechanism-based LSTM network, Automation in Construction 125 (2021) 103647.
[40] Z. Gui, Y. Sun, L. Yang, et al., LSI-LSTM: an attention-aware LSTM for real-time driving destination prediction by considering location semantics and location importance of trajectory points, Neurocomputing 440 (2021) 72–88.
[41] Y.L. He, L. Chen, Y. Gao, et al., Novel double-layer bidirectional LSTM network with improved attention mechanism for predicting energy consumption, ISA Transactions 127 (2022) 350–360.
[42] W. Guo, C. Wu, Z. Ding, et al., Prediction of surface roughness based on a hybrid feature selection method and long short-term memory network in grinding, The International Journal of Advanced Manufacturing Technology 112 (2021) 2853–2871.

Subject index

G

H

I

Q

R

S

T

U

Printed and bound by CPI Group (UK) Ltd, Croydon, CR0 4YY
14/09/2026
02216409-0001